Antibiotics

David M. Shlaes

Antibiotics

The Perfect Storm

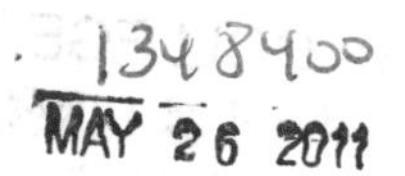

David M. Shlaes
Anti-infectives Consulting, LLC
Montauk Avenue 219
06378 Stonington
CT, USA
shlaes.david@earthlink.net

ISBN 978-90-481-9056-0 e-ISBN 978-90-481-9057-7
DOI 10.1007/978-90-481-9057-7
Springer Dordrecht Heidelberg London New York

Library of Congress Control Number: 2010934450

Printed on acid-free paper

Springer is part of Springer Science+Business Media (www.springer.com)

To Jan for years of hard work, patience,
understanding, love and support.

David M. Shlaes MD, PhD has had a 30-year career in anti-infectives spanning academia and industry with a long-standing scientific interest in antimicrobial resistance. In 1991 he was appointed Professor of Medicine at Case Western Reserve University. In 1996, Dr. Shlaes became vice president for Infectious Diseases at Wyeth Research for 6 years, assuming responsibility for the strategic direction for infectious diseases within Wyeth. In 1998 Dr. Shlaes was the cover feature in the April issue of *Business Week* dedicated to antibiotics research. In 2002, Dr. Shlaes became executive vice president, Research and Development for Idenix, Pharmaceuticals, a company located in Cambridge, MA, focused on the discovery and development of antivirals. In 2005, he left Idenix to form a consulting company for the Pharmaceutical Industry (Anti-Infectives Consulting, LLC). He was recently an independent director for Novexel, S.A, an anti-infectives biotech in Paris that was just sold to Astra-Zeneca. He consults for a number of other anti-infective focused biotechs and frequently works with VC firms in the evaluation of anti-infective companies.

Contents

Chapter 1
The Perfect Storm

On the antibiotics front, the weather has occasionally been bad ever since I can remember. But things have never been as bad as they are today. Bacteria are becoming resistant to the point where none of our available antibiotics work for some of the infections that confront patients and physicians in hospitals around the US and the around world. For these patients we are slipping back in time to a pre-antibiotic era where we have little to offer but comfort for diseases which we have been easily able to cure over the last 50 years.

But in answer to resistance, our antibiotic pipeline is all but dry and the situation is deteriorating. The science of discovering new antibiotics is exceedingly challenging and the economics of antibiotics are becoming less and less favorable. The regulatory agencies like the FDA are contributing to the problem with a constant barrage of clinical trial requirements that make it harder, slower and more costly to develop antibiotics. The pharmaceutical industry, under extraordinary financial pressures, is consolidating at historic rates leaving fewer and fewer large companies standing. The antibiotic market is not as promising as markets for treatment of chronic diseases like high cholesterol or chronic depression or high blood pressure. For those diseases which we cannot cure, the drugs must be taken for long periods of time, frequently for a lifetime. Antibiotics, which actually cure disease, are only taken for days or weeks.

In response to all these pressures, of the large pharmaceutical companies still extant, fewer and fewer are remaining active in antibiotic research. If this isn't The Perfect Storm, I don't know what is.

There are possible solutions to this conundrum. They include incentives to companies for antibiotic research and development, deconsolidation within the pharmaceutical industry and affecting a more balanced approach within the FDA that assures us the ability to develop new antibiotics for resistant bacteria.

Basically, we either invest now with organization, balance and money or we will pay with our lives. Antibiotics are going the way of the dodo but bacterial pathogens are with us for the duration.

I feel like I've been fighting this battle all my life. When I finished training and started my career in academic medicine, I wanted to study antibiotic resistance in bacteria. I felt that if we understood how resistance spread, we could stop it. If we

D.M. Shlaes, *Antibiotics*, DOI 10.1007/978-90-481-9057-7_1,

could understand the mechanisms by which the bacteria become resistant in detail we could find ways around them with new antibiotics.

My first job put me in an ideal position to do that. I was director of the clinical microbiology laboratory at the Veterans Affairs Medical Center in Cleveland and I worked in the infectious diseases service. So I was seeing patients and their infecting bacterial pathogens every day. The problem was that research doesn't pay for itself. Someone has to provide the money to support the work. The Veterans Administration has a wonderful research support system that allowed me to get started and funded me virtually constantly for the 16 years I worked there. I will always be grateful to the VA for their constant support. Their approach should be contrasted to that at the National Institutes of Health (NIH) where there was, for many years, a bias against funding antibiotic resistance research. They felt that antibiotic resistance was not important or at least that it was somehow not good science. If it was important, the pharmaceutical industry and not the NIH should fund it. The pharmaceutical industry was not so interested in resistance in those days either since it usually did not do any good for marketing their products. I submitted grant requests to the NIH – but they were mostly not successful. It wasn't just me. There were only a very few of us in the United States working on antibiotic resistance and we all knew each other. One reason there were so few of us was that it was so difficult to obtain support for our research. When we got together we realized that we all had the same problem. In 1986 we met and began to investigate the NIH and their funding practices as far as antibiotic resistance was concerned. We confirmed that they did not fund much research in the area and requested a meeting with them. We found that one major reason they did not fund resistance research was that they had very few people working on reviewing grant proposals who were even familiar with antibiotic resistance. Since the NIH relied entirely on the grant reviews provided by their reviewers, resistance researchers were left out. We actually held several small workshops with the NIH over the next 5 or 6 years. One of our main recommendations to the NIH was to establish a special peer group to review grant requests in the area of antibiotic resistance. As a result of our efforts, we were the subjects of an article in the prestigious journal Science that referred to us as disgruntled scientists. Finally, 20 years after our group first met, the NIH established the very peer review group we recommended. Any discrimination against resistance research in the NIH seems to be gone, but grant money is still very hard to come by. The STAAR Act (Strategies to Address Antimicrobial Resistance) is currently under consideration in congress. It seeks to strengthen the surveillance, prevention and control and research efforts in antibiotic resistance in part through additional NIH funding.

In 1996 I took the leap to industry where I became Vice President for Infectious Diseases at Wyeth Pharmaceuticals. I moved to industry for a variety of reasons. One of them was that I learned that if I truly wanted us to have new antibiotics active against resistant bacteria, I needed to be in industry and not academia. And I truly wanted us to have new antibiotics! At Wyeth, as the infectious diseases therapeutic area head, my team (BIG) and I were responsible for discovering and ultimately for developing and even marketing Wyeth's antibiotics. Tigecycline (Tygacil), launched in 2005, was partly the result of those efforts. Getting tigecycline through the FDA required another battle that I describe in Chapter 4. At least no one was saying I was disgruntled. I stayed at Wyeth until 2002 when I went to a small biotech working on antiviral drugs (where drugs against resistant viruses are important, too). There I was responsible for drug discovery, manufacturing, pharmacology and toxicology. In 2005 I became a consultant. I currently work with a number of companies, mostly small, who are trying to discover and develop new antibiotics for resistant bacterial infections. I still participate in research when I can and I try to stay involved in the ever-changing landscape of FDA-industry relations. I still dream of bringing more new antibiotics that work against resistant bacteria to patients and their physicians. But that dream is becoming harder and harder to realize.

I chose the specialty of infectious diseases because, during my training in the late 1970s, I loved being able to identify the cause of a patient's infection, choose the right antibiotic and watch my patient get better in a matter of hours to days. Of course, it didn't always work that way. Once I was called to the burn unit at the county hospital where I was training in infectious diseases. One of the patients was a severely ill middle aged gentleman with an infection caused by a Gram-negative bacterium. When I arrived, the patient had a high fever and was on a ventilator (breathing machine). Tubes were everywhere. He had severe second and third degree burns on his face, chest and abdomen that were draining a greenish-white material. His doctors had already sent a culture of this material to the microbiology laboratory. The report was in his chart and was the reason his doctors called for help. The bacterium that grew in culture was a Gram-negative organism called *Pseudomonas aeruginosa* and the green pigment it produces was responsible for the color of the purulent drainage from the patient's burn. The microbiology lab had tested the isolate for susceptibility to about 10 different antibiotics. As I went down the list, all I saw was R (for resistant) except for one S (for susceptible). This indicates that the patient's *Pseudomonas* was only killed by one of the antibiotics tested in the lab. The antibiotic with the S was colistin, discovered in 1947, and something I had never used and that was essentially never used even back in the 1970s. The doctors in the burn unit were, of course, already treating him with antibiotics, but the patient had not improved. While I was on the unit, I got a call from the lab informing me that the patient's blood was also growing a Gram-negative bacterium (later shown to be the same *Pseudomonas aeruginosa*). I called my supervisor, explained the situation, and we agreed that we would start the patient on colistin. My supervisor had to help me figure out how much to give and how often because this antibiotic essentially had not been used since the 1960s. The antibiotic was known to be toxic to nerves and to the kidney and because it had been developed so long ago, we didn't really know how well it would work.

I came back to see the patient the next day. He still had fever. His wound was unchanged. The bacterium was sensitive to the antibiotic we had prescribed. I obtained additional cultures of the wound and blood. The surgeons continued to debride the wound as much as possible. There seemed nothing else to do. Within a few days, the new blood and wound cultures were still growing the same organism, still sensitive to the drug we were using to treat him, but he was no better. His kidney function began to deteriorate and he was dead within a week. This was not the only similar infection that I saw on the burn unit that year. Most, but happily not all, had a similar outcome. Today, physicians are once again forced to use this old antibiotic, colistin, to treat infections caused by bacteria resistant to everything else including all the antibiotics introduced since the late 1970s. There are new studies being conducted to determine whether colistin really works or not and to quantify its level of toxicity. One recent study from a US military hospital where Iraq war veterans have highly antibiotic resistant infections and where they are using colistin fairly often recently reported that 21% of their patients had to stop colistin because of kidney toxicity. In a few years, maybe we'll know the whole story.

I first arrived at the Cleveland Veterans Administration Medical Center in 1980. I had just finished my training in internal medicine and in infectious diseases so this was my first real job. A few years later, I was called to the surgical intensive care unit to see a patient with pneumonia. He was another middle-aged man who had undergone a coronary artery bypass about a week prior to my visit. He was still on a ventilator, but could communicate with me. When I examined him, he had fever and his chest was full of pneumonia on one side. His physicians had obtained cultures of the secretions coming from his pneumonia through his breathing tube. The microbiology report was just like the one from the county hospital a few years earlier – all the antibiotics were R. Colistin was not tested, but I was wary of using it again. I knew that Merck Sharpe and Dohme had a promising new antibiotic in the clinical testing stage and that I might be able to get hold of some for this patient. I called, filled out some (!!) paperwork, obtained informed consent from the patient and discussed the situation with his family who were supportive. We received a shipment of this new antibiotic within 24 hours. I started my patient on the new drug. The bacterium responsible for his pneumonia was susceptible to the new antibiotic. Within a few days, his fever was down and within a week he was able to come off the ventilator. As it turns out, the surgical unit had several cases like this all of which we were able to treat with the experimental antibiotic. We only lost one. The drug, imipenem-cilastatin (Primaxin), was eventually approved by the FDA and is still marketed today.

These true stories illustrate several important points. The first case shows what happens when we don't have antibiotics that work against resistant organisms. We are seeing more and more examples of this as we will see throughout this book. An infection with bacteria resistant to all available antibiotics was a rare occurrence during the "golden years" of antibiotic discovery from 1955 to 1985. During those years, there was almost always something new and active against resistant strains in late stage clinical trials or entering the marketplace. For the really old antibiotics, those approved before, say 1980, we are sometimes not so sure of either how well they work or how toxic they are. Many were never studied as carefully as we study antibiotics today.

The second case illustrates that antibiotics are truly miracle drugs that cure disease and save lives. It also exemplifies a more typical scenario during the golden years where there was always a new antibiotic in the pipeline that we could use to treat patients with antibiotic-resistant infections. Of course we want to try and prevent or slow the emergence of resistance as much as possible, but we also want to have new antibiotics in the pipeline so that we are not faced with infections we can't treat. Currently, we are not doing as well as we could on the first front and we are failing miserably on the second.

About 2 million people acquire an infection during their hospitalization in the United States each year. 90,000 of them die. Many of these patients will have infections caused by resistant bacteria, and the resistance will contribute to their deaths. Antibiotic resistant infections increase death rates by 50–100% in general. Why? Partly because doctors don't suspect the resistance and will use the wrong antibiotic initially. In fact, if the doctors don't or can't obtain a sample of the infected material for culture, they won't know if the organism is susceptible to the antibiotic they are using or not. And frequently, a culture is not obtained, and the physician plays the odds based on what is known about resistant bacteria in his or her particular community or hospital.

Bacteria in our communities are also becoming more resistant. The latest example is staph (*Staphylococcus aureus*). Staph are bacteria that cause everything from minor skin infections like impetigo in children to minor abscesses to very serious skin infections (so called flesh eating infections) to bloodstream and heart infections which are fatal 30–40% of the time. We have had sequential epidemics of ever more antibiotic-resistant staph infections in our hospitals and communities since the 1950s.

When penicillin was first discovered, even before it was tested in humans, the first resistant bacterial strains were discovered. Like many antibiotics, penicillin is produced by microorganisms found in the soil. They produce antibiotics to help them compete with other organisms in their ecological niche. These organisms have, in turn, evolved ways to resist the chemical weapons of their competitors. So, that we could find bacteria resistant to penicillin even before we started to use it should not be so surprising. When penicillin was finally brought to the public marketplace after World War II, almost all strains of staph were still killed by the drug. By the 1950s, the majority of hospital strains of staph were resistant. By the 1970s both hospital and community strains were equally resistant. Luckily, by then, we had tetracycline, erythromycin and other antibiotics to use since penicillin was no longer effective.

Today, staph that are resistant to many of our most useful antibiotics (called MRSA) cause about 60–70% of staph infections in communities all over the US. The MRSA causing infections in our communities, unlike those in our hospitals, are usually still sensitive to a few of the older antibiotics like bactrim (which contains a sulfa drug) or sometimes tetracycline. But neither bactrim nor tetracycline, since they were approved many years ago, was carefully studied in the treatment of these kinds of staph infections. For these infections, there is only a single oral (a pill) antibiotic, linezolid (Zyvox), which has been well studied and been shown to be safe and effective. Even then, linezolid can be toxic if it is used for longer courses of therapy such as for serious bone infections. Because linezolid is so expensive, physicians still frequently use bactrim and tetracycline to treat resistant staph infections, even though we don't know how well they work. Also, among older folks, bactrim allergies are common and can result in serious reactions. Otherwise, the only choice is to use intravenous antibiotics. The modern intravenous antibiotics have been much better studied so we are more comfortable with the fact that they will work and we know their toxicities very well. But, to get an IV antibiotic, you would either go to a hospital (bad idea if you can avoid it) or get your IV therapy at home. Our choices are limited. For the past 10 years I have been asking myself, what will happen if staph becomes resistant to the remaining few antibiotics that still work? Will we have something to use then? I'm not so sure . . .And in fact, MRSA have been creeping towards being more resistant to even our remaining IV antibiotics. This day may come sooner rather than later.

With no exceptions, like it or not, the antibiotics we have available to us today came to us fully or at least partly through the efforts of the pharmaceutical industry. The ability of the industry to continue to provide us with new and more active antibiotics is rapidly disappearing. Our ability to discover new antibiotics is decreasing because it is becoming harder and harder to identify non-toxic molecules that will kill bacteria. All the obvious antibiotics have apparently already been discovered. Everything that will come later will come harder.

In this background of growing resistance and more difficult science, the pharmaceutical industry continues to consolidate. By 2003, the industry had undergone a consolidation of more than 90% over the prior 20 years. Since then, Sanofi has merged with Aventis to form Sanofi Aventis, Pfizer has purchased Wyeth and Merck has purchased Schering-Plough. This will result in over a 95% consolidation during the past three decades. This consolidation results in fewer and fewer researchers working on new antibiotics.

Because, since the late 1990s, the pharmaceutical industry has been questioning the market value of new antibiotics, more and more companies have simply abandoned this field of research entirely. After all, if you are selling pharmaceuticals, would you rather sell a Z-pack of azithromycin (Zithromax) to be taken over 5 days or would you rather sell a Lipitor for high cholesterol that has to be taken daily forever? Of course, in my view, this is an oversimplification that has recently been proved wrong by linezolid. We will explore this in greater detail in a later chapter. Nevertheless, among the few large companies that are left, even fewer are actively looking for new antibiotics.

Contributing to the negative view of antibiotics by the pharmaceutical industry are the conscientious efforts by physicians to prevent resistance. Most physicians believe (rightly in my view) that antibiotic resistance is more likely to appear of the antibiotic is heavily used. If a new antibiotic comes out which is active against resistant bacteria, physicians tend to try and reserve it for use only when truly necessary. Don't get me wrong – I'm not saying this is a bad idea. While this may be good for public health, it is not good for the revenues of the pharmaceutical companies that want to sell the antibiotic.

Finally, there is the FDA and other regulatory agencies. The Food and Drug Administration is charged with ensuring the efficacy and safety of drugs allowed on the US market. In their good faith efforts to do their job, the FDA has led us to a situation where their new requirements for studying antibiotics in order to obtain approval will only further alienate the industry from antibiotic research. These tough requirements may be based on good science, but they render the clinical trials at best impractical and at worst infeasible. Their trial requirements have essentially removed large portions of the antibiotic market from the US for the foreseeable future. As we will see later, introducing a new antibiotic for mild bacterial infections like sinusitis, bronchitis and ear infections to the US market has now become virtually impossible. Even for a more serious infection like pneumonia, the development of new antibiotics has become much more difficult and expensive if not impossible. There may be sound scientific reasons for questioning the benefit of antibiotics for some of these infections, but the industry just sees a black hole in their bottom line for antibiotics.

In the following chapters, I will explore the good and the bad about antibiotics. I will discuss the problem of bacterial resistance, how this happens and the bacterial threats we face today and those on the horizon. I will dissect the history and present state of both the pharmaceutical industry and the FDA in the realm of antibiotics. Finally, I will discuss potential solutions for the industry, the FDA, for researchers and for us as a society, which might allow us to continue to have antibiotics available for use when we need them most.

Chapter 2
The Miracle

The history of antibiotics goes back to mercury and bismuth, heavy metals which are toxic to people, but in correct doses, were more toxic to the organism that causes syphilis, *Treponema pallidum*. Mercury as a therapy for syphilis was first discussed back in the 1400s, but the heavy metals were not widely used until the end of the nineteenth century. Whether they were efficacious or not is not entirely clear since no systematic studies like those we use today were carried out in those times. But there is good reason to believe that this therapy worked at least to some extent. Of course, as all of us who like to eat fish know, mercury is also rather toxic to people.

Salvarsan, an arsenical compound, was discovered by Paul Ehrlich and his co-workers in 1908 and was marketed in 1910. The arsenicals were discovered based on their activity against parasitic microorganisms, but only were developed as antibiotics when they were active in a rabbit model of syphilis. Salvarsan was dubbed Ehrlich's magic bullet. He won the Nobel Prize for his discovery in 1908. Again, the kinds of clinical studies we are accustomed to today were not undertaken, but various testimonials from "miracle" cures were used to demonstrate the efficacy of Salvarsan. In retrospect, there are good reasons to believe that the drug had some efficacy in the treatment of human syphilis. The use of the drug was limited by its toxicity – it was based on arsenic after all.

These early efforts point out one of the great challenges of finding good antibiotics. We are looking for something that will kill the microorganism but that will leave its host (us) unharmed. We want a very specific toxin. This is not so easy. The later discoveries of sulfonamides and then penicillin and its relatives misled many into believing that finding such specific microbe killers that were perfectly safe for humans might be less difficult than it is.

The discovery of the sulfonamide antibiotics by Bayer in Germany in 1932 was the next great leap forward for antibiotics. The first such drug was called Prontosil. It was actually a prodrug because it relied on the human body to metabolize Prontosil into its active form, sulfanilamide. It was first synthesized by Bayer chemists Josef Klarer and Fritz Mietzsch. Prontosil was tested and found effective against some important bacterial infections in mice by Gerhard Domagk, who subsequently received the 1939 Nobel Prize in Medicine. Its utility was dramatized in an article in Time Magazine in 1936.

D.M. Shlaes, *Antibiotics*, DOI 10.1007/978-90-481-9057-7_2,

> Last month Mrs. Franklin D. Roosevelt, who loves few things better than a big family feast, gave up Thanksgiving dinner at Hyde Park to rush to Boston where Son Franklin Jr. lay abed with what was described to the press as "sinus trouble." The young man did have infected sinuses, and he was in the capable, Republican hands of Dr. George Loring Tobey Jr., a fashionable and crackerjack Boston ear, nose & throat specialist. He also had a graver affliction, septic sore throat, and there was danger that the Streptococcus haemolyticus might get into his blood stream. Once there the germs might destroy the red cells in his blood. In such a situation, a rich and robust Harvard crewman is no safer from death than anybody else.
>
> Not until last week, when his mother and his fiancée, Ethel du Pont, went home, was Franklin Jr. out of danger and fit for Dr. Tobey to operate on his infected right antrum (in the cheek) and ethmoid sinuses (in the brow). Simultaneously, Dr. Tobey let it be known that his notable young patient had been pulled through his crisis by a notable new drug.
>
> When Franklin Roosevelt's throat grew swollen and raw and his temperature rose to a portentous degree. Dr. Tobey gave him hypodermic injections of Prontosil, made him swallow tablets of a modification named Prontylin. Under its influence, young Roosevelt rallied at once, thus providing an auspicious introduction for a product about which U. S. doctors and laymen have known little.
>
> The drug which cured young Roosevelt seems to be a specific cure for all streptococcic infections—septic sore throat, childbed fever, postabortal septicemia. It has helped to cure cases of peritonitis due to ruptured appendix, perforated stomach ulcer or gallbladder. It has been effective in postoperative wounds, endocarditis, suppurative mastoiditis, and tonsillitis. Some cases of erysipelas (also a streptococcic infection) have yielded to Prontosil medication. The drug also has ameliorated severe cases of carbuncles and cellulitis due to staphylococcus, a different kind of germ.

Figure 2.1 below shows the mortality rates at Cook County Hospital in Chicago from erysipelas, a serious skin infection caused by *Streptococcus pyogenes*, the same organism that causes strep throat. Clearly, for this infection, the use of sulfonamides saved lives. The same thing was shown for pneumonia as shown in the table from a 1939 article below.

Dagenan was a sulfonamide antibiotic. As you can see from Table 2.1, the mortality rate for treated patients ranges from 6 to 17.6% while for untreated controls it

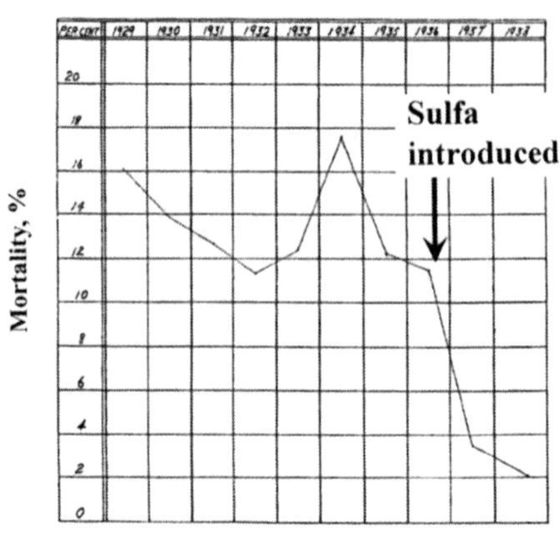

Fig. 2.1 Mortality rates of erysipelas at Cook County Hospital in Chicago, Illinois, from 1929 to 1938. Sulfonamides (sulfa) became generally available in 1936. From Spellberg et al., 2009 with permission

Table 2.1 Pneumonia and mortality 1938

Mortality			
Controls	Controls with positive blood culture	Dagenan treated	Dagenan treated with positive blood cultures
23%	50%	6%	17.60%

ranged from 23 to 50%. There is no doubt that in pneumonia, sulfonamides saved lives and did so in a dramatic way.

Penicillin had even more dramatic effects.

The first patients to be treated with penicillin were in England during 1940. The team of Florey, Chain and Heatley at Oxford University was producing the penicillin. There they had developed a kind of Rube Goldberg apparatus to purify the drug. Florey and Chain went on to win the Nobel Prize for their efforts to characterize, manufacture and test penicillin. There was just a tiny amount of drug such that, because the drug was excreted unchanged in the urine, the urine of the treated patients was collected and the penicillin repurified for use again. One patient, Mr. Alexander, was being slowly "eaten" with a mixed staph and strep infection. He had pus everywhere. One eye had to be removed. His shoulder and lung became infected. Sulfonamides were not helping. He agreed to receive penicillin. Within 24 h, there was dramatic improvement. His temperature returned to normal, his lesions improved and his appetite returned. Five days later, treatment was stopped because they were running out of penicillin and they had not yet had time to repurify the penicillin from the patient's urine. Of course, no one knew how long one should treat in any case. This is still a problem today I might add. Unfortunately, 10 days later, Mr. Alexander had a relapse of his lung infection and there was no more penicillin to be had. He died a month later.

In 18 months of testing in something like five or six patients, 4 million units of penicillin had been used. This would be roughly equivalent to one sixth of the minimum daily dose we would use for a single patient with pneumonia today.

The first American patient was treated at Yale in March of 1942. Mrs. Miller had staph infection of her bloodstream and was dying. One of the penicillin researchers was hospitalized nearby with a viral infection. Mrs. Miller's doctors asked him if he would be able to procur penicillin for Mrs. Miller. He did. From Merck. She lived. Her temperature went from 105.5°F before treatment to normal in less than 24 h.

My father was a physician who practiced internal medicine in Chicago his entire life. He didn't retire until he was 83 years old and was unable to walk independently anymore. He always told a fascinating story about my aunt who had juvenile onset diabetes. Diabetes, for reasons still not very clear, makes people more susceptible to infections. My Dad was doing his internship in New York City in 1944. My aunt developed a breast abscess with staph. The staph then went on to invade her bloodstream. She was septic, on the verge of death. In those days, the military had essentially the entire supply of penicillin (it did not become available to the public until after the war in 1945), but my Dad knew that this was her only chance. He called the public health office in New York who put him in touch with the military from which he was able to procure a small supply of penicillin. He administered several hundred thousand units – a large dose at the time. Within 24 h my aunt was sitting up in bed eating and talking. Her fever was gone in a few days. Based on this experience and others, he always believed in the miracle of antibiotics.

In Table 2.2 we can see data from the pre-antibiotic years compiled by the Infectious Diseases Society of America. They compare controls where treatment might be expected to be ineffective with sulfonamide and penicillin treated patients with serious skin infections. Possibly because this group included patients with infections that are less serious than erysipelas, the cure rates with ineffective therapy are higher than we saw when we just looked at erysipelas earlier. Nevertheless, both antibiotics work and penicillin seems to work even better than sulfonamides.

Even today, most physicians will be able to relate a story something like my father's. Of course, most do not have to procure their supply of antibiotic from the military, but they will all be able to tell you about a patient on death's door who comes back from the brink after a few doses of an antibiotic.

Table 2.2 Treatment of erysipelas

Penicillin vs. erysipelas/cellulitis	Ineffective therapy	Sulfonamide	Penicillin
Patients cured (%)	1520/2294 (66)	1423/1573 (91)	196/200 (98)

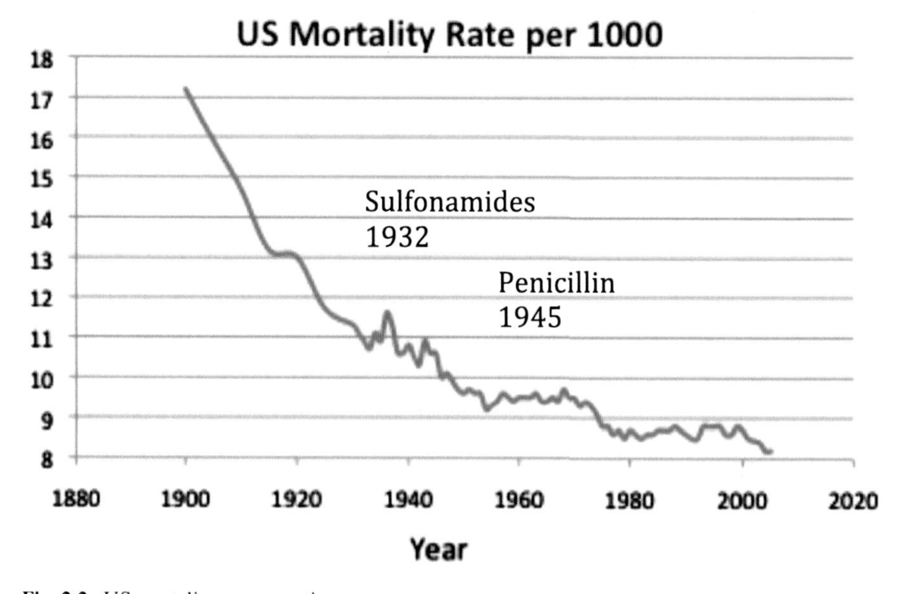

Fig. 2.2 US mortality rate over time

In looking at a chart of US mortality rates going back 125 years (Fig. 2.2), we can see that lots of things contributed to lower mortality besides antibiotics. Better sewers, clean water supplies, better public health measures, and vaccines all contributed to the lowering of mortality rates over the years. The chart above shows where sulfonamides and penicillin were introduced to the market in 1932 and 1943. Many experts feel that the introduction of antibiotics made an important contribution to lowering mortality rates as well. For example, during the last few decades, taking all comers with community-acquired pneumonia of varying degrees of severity, the overall mortality is around 3%.

Today, the miracle of antibiotics is being threatened on all fronts. The bacteria are becoming steadily more and more resistant even to our newest antibiotics. Our ability to discover new antibiotics is restrained by the more difficult science. It seems that nature has already discovered all the best antibiotics. We are further hindered by the lack of researchers in the area. The years of lack of NIH support for antibiotic research have taken their toll. The pharmaceutical industry, our primary source of antibiotics, has been consolidating over the last three decades. The fewer companies there are, the fewer researchers there are. In addition, individual companies have abandoned antibiotics as a field of research. This further contributes to our lack of effort in discovering new antibiotics at a time when discovering them is more difficult. For some patients, we are already back to the year 1900 in terms of available, effective antibiotics. That number of patients can only rise over the next 10 years.

Chapter 3
Resistance

The Basics

Bacteria resistant to antibiotics existed long before antibiotics were even a twinkle in the eye of the pharmaceutical industry. Most of the antibiotics on the market today, including penicillin, erythromycin, tetracycline and all their relatives started as byproducts of the metabolism of microorganisms in soil, on plants or in our oceans. Antibiotics are part of the never-ending competition among species for survival or dominance within an ecological niche.

To start with, if you are a fungus producing a toxic metabolite, say penicillin, you better be able to resist your own toxin. Microorganisms have evolved an entire bag of tricks to evade the toxic effects of their own antibiotics and those of other competitor organisms. Each toxin produced has some particular target. For example, penicillin targets the enzymes that make the cell wall of bacteria. Fungi, including the mold that produces penicillin, *Penicillium*, do not have the same type of cell wall and therefore do not even have the enzymes targeted by penicillin. Other tricks to avoid these poisons include pumps which literally pump the toxin back out of the bacterial cell faster than it diffuses in; or enzymes that either degrade the antibiotic or modify it in a way that it can no longer work; or even to have an alternative pathway for making your cell wall such that your pathway cannot be blocked by antibiotics that target enzymes in the normal pathway. Some organisms modify the target of the antibiotic such that the antibiotic can no longer bind. It is important to understand that all of these bacterial tools to deal with antibiotics were present for eons before man started using antibiotics to cure disease.

Enter the age of antibiotics. You are a bacterium. You can divide and reproduce every 20–30 min. You can easily exchange genetic material with other bacteria. You are invincible! This principle is the basis for my belief in the "you use it you lose it" theory of antibiotics. If you introduce a new antibiotic into use in people or animals or for crops or any combination of these, you immediately apply pressure on the bacterial populations. All living things live in some kind of relationship with bacteria. Humans (and animals and plants) have enormous populations of bacteria living on their surface and in their guts. These bacteria serve a useful purpose – they crowd out the bad actors, they help digest some foods, etc. As soon as you add

D.M. Shlaes, *Antibiotics*, DOI 10.1007/978-90-481-9057-7_3,

large quantities of an antibiotic to the mix, everything is turned upside down. We kill the bad guys (we hope) but we also kill a lot of the good guys. And everybody is trying to survive. If, in that population of bacteria, there are just a tiny number that are resistant to the antibiotic in question (and there frequently are), when you apply the antibiotic those few will survive. With time, they will multiply and may come to dominate the niche during antibiotic therapy and even afterwards in some cases. These resistant bacteria may be able to share the mutation or the gene causing resistance with other bacteria. They may have acquired some gene coding for resistance to the antibiotic in question eons ago, but this may be the first time it has actually been needed for a major assault. This foreign gene may also be capable of transfer to other bacteria. How fast all this happens and how quickly it spreads is very variable. So some antibiotics were on the market for decades before we discovered significant resistance to them while resistance plagued others before they were ever manufactured in large scale.

The other point I am making is that resistance may be more likely to arise out of collateral damage to the friendly bacteria we live with all the time rather than through our efforts to knock off the particular organism causing the infection we are treating. The friendlies can mix with the pathogens on our skin or in our gut and transfer their resistance genes. An interesting possible example of this is *Staphylococcus sciuri*, a species of staph that is essentially not pathogenic (rarely causes disease) and is found more often colonizing the skin of animals. It seems to be the origin of the gene that is responsible for methicillin resistance in *Staphylococcus aureus* (MRSA). *S. sciuri* with this gene might have been selected when penicillin began to be used in veterinary medicine back in the late 1940s. This is now coming back to haunt us with a vengeance with the highly resistant and virulent MRSA strains we currently face in our hospitals and communities.

Antibiotics for Animals and Crops Lead to Resistance for People

And speaking of collateral damage, at least half of all the antibiotic use in the United States is for animals and crops. This has been a controversial topic for over 30 years. A 1998 report from the prestigious Institute of Medicine of the National Academy of Sciences noted that about 4 million pounds of antibiotics were used to treat sick farm animals and another 16 million pounds were used as growth promotants (low doses of antibiotics usually included in animal feed) for animals every year. Nobody knows to this day why low doses of antibiotics make animals grow faster, but its true. They grow about 4–8% faster than animals not fed on low dose antibiotics. Of course, you then have to add the additional 300,000 pounds of antibiotic pesticides, mostly tetracycline and streptomycin, which are used every year on crops. This total is roughly equal to the total antibiotic use for the treatment of people each year.

E. coli is a common bacteria colonizing our intestinal tract, but also causing urinary tract infections. It can cause severe sepsis and meningitis (inflammation of the membrane surrounding the brain and spinal cord) in newborns, especially the premature infants. It also is a common cause of diarrhea, especially that disease of travelers we call *la tourista*. More rarely, certain strains cause a very severe diarrhea sometimes associated with kidney failure – the famous *E. coli* OH157 of undercooked burger fame. It tends to remain rather susceptible to antibiotics globally. But, about 60% of isolates causing infections in patients in the US are resistant to tetracycline. Of course, tetracycline is still used in humans as well as animals and for crops. But surprisingly, around 20% of strains were also resistant to streptomycin even though this drug is hardly ever used for treatment of people anymore. There is good reason to believe that part of this is due to streptomycin use for crops.

A more dramatic example comes from an organism called *Enterococcus*. Starting in 1989, US hospitals have experienced an outbreak of *Enterococcus* resistant to what was our last line antibiotic at the time, vancomycin (they were frequently already resistant to everything else). About 50% of US hospital strains are now resistant to all but our latest two last line antibiotics, daptomycin and linezolid. During the late 1980s and early 1990s, there was no daptomycin and no linezolid for infected patients. There was nothing that worked. It turns out that the resistance seems to have been derived from animals in Europe where they were fed low concentrations of an antibiotic called avoparcin to promote growth. Avoparcin is closely related to vancomycin and the vancomycin-resistance genes found in enterococci infecting people are essentially identical to those found in animals fed avoparcin. Although this is one dramatic and threatening example of the transmission of antibiotic resistance from animals to humans, there are lots of others.

Salmonella is a type of bacteria that causes diarrheal disease in humans. It is frequently carried by animals but can cause diarrheal disease in them as well. Many of the recent food-borne outbreaks in the US that resulted in large recalls (peanut butter for example) were caused by *Salmonella*. *Salmonella typhimurium* DT104 was first isolated from a human stool specimen in England in 1984 and was resistant to the antibiotics commonly used to treat such infections including bactrim and tetracycline. It remained relatively rare until around 1990. In 1993 the strain became epidemic in Europe and in the US. It rapidly acquired resistance to other antibiotics including the quinolones and the modern cephalosporin antibiotics (related to the penicillins). It is clear that this strain can be transmitted back and forth between humans and animals. Animals treated with antibiotics are more likely to harbor the strain. The headline from a local Colorado newspaper below (Fig. 3.1) refers to King Soopers of Denver (a subsidiary of Kroger Foods) who recalled over 466,000 pounds of ground beef (about 1000 cows' worth of meat) for DT104 contamination causing illness in several states. DT104 is now found throughout the world. In one recent outbreak, almost 50% of infected patients were hospitalized and 10% died of their infection. The strongest risk factor for infection with this strain in humans is the recent use of any of the antibiotics to which the strain is resistant. For these multiply resistant strains of *Salmonella*, we have a single, reliable class of antibiotics left in our armamentarium, the carbapenems that includes

Contaminated beef recalled from King Soopers, City Market (reprinted with permission)

Friday, July 24, 2009

About a half-million pounds of King Soopers ground beef is being recalled by King Soopers of Denver for possible Salmonella contamination Colorado Department of Public Health and Environment officials announced Thursday. The ground beef was sold statewide at King Soopers and City Market retail grocery stores. Identified as Salmonella typhimurium DT104, it is a strain resistant to many antibiotics prescribed for treatment which can increase risk of hospitalization, or possible treatment failure in infected individuals. Although the product may be no longer available in stores, state health officials urge consumers who may have purchased the product between May 23 and June 23, at local King Soopers and City Market stores, to check their freezers for any product and discard it or return it to the place of purchase.

Fig. 3.1 Headline article from a local Colorado Newspaper with permission

imipenem (from Chapter 1) among others. You will see that this will be a repeated theme in this chapter.

An underappreciated use of antibiotics occurs in fish farming. Here, it is exceedingly difficult to quantify the antibiotics used, but the quantities, and potentially the human health consequences might be no less important than antibiotic use in other animals and in crops. A recent publication from Scandinavian scientists highlighted the problem. They noted that one study of ready to eat shrimp (13 brands in four different countries) showed that 42% of the bacteria recovered from the shrimp were antibiotic resistant and that 81% of the bacteria isolated were human pathogens including *E. coli* and staph. Clearly, antibiotic-resistant organisms from farmed fish pose a risk of worldwide spread of antibiotic resistant organisms and their resistance genes.

Antibiotic resistance in bacteria is often carried on segments of DNA that can jump from one organism to another. One recent report documented that these DNA segments carrying multiple antibiotic-resistance genes were identical in *E. coli* from humans and animals and in *Salmonella* from humans and animals.

Our environment in general is becoming polluted with antibiotics. Runoff from farms and from our own sewers (yes, antibiotics are excreted in urine and in feces) exposes more and more of the worlds bacteria to low concentrations of antibiotics. The world, in the words of Julian Davies, one of the original and most important researchers in antibiotic resistance, is a dilute solution of antibiotics. This will continue to select for antibiotic resistance among environmental organisms. A *Pseudomonas* with a new gene causing resistance to our last line antibiotics for these bacteria and associated with multiple other antibiotic resistance genes was recently fished out of the Seine River in Paris. In a recent study of *E. coli* from recreational beaches and private drinking water in Canada, of a total of about 15,000 isolates, 142 were highly resistant to multiple antibiotics. This represents a small percentage but shows the risk we are taking.

Europe has already banned the use of any antibiotic related to those used in human health for non-therapeutic use in animals. In the US, congress is once again considering a similar ban (HR 1549). Given that this discussion has been ongoing for over 30 years, one wonders what everyone has been thinking. Of course the bill is strongly opposed by the beef and poultry industry.

The issue of antibiotic use animals as treatment or prevention of infection is also controversial. For example, one of the antibiotics used in this way is a fluoroquinolone closely related to antibiotics used in people. It is clear that resistance to the human antibiotics like ciprofloxacin can be selected by therapy in animals and that these resistant strains or their resistance genes can be transmitted to people. The FDA has continued to allow therapeutic use of these antibiotics in animals, and has been doing so for almost 15 years while monitoring data. As a first step, the World Health Organization has recently ranked antibiotics according to their importance in human health. Their top priorities are the fluoroquinolones, the macrolides (like erythromycin and azithromycin) and the modern cephalosporins. The idea would be to identify antibiotics that could be used in animals that would be so unrealated to those used in human health that resistance to human antibiotics would not be selected by the veterinary antibiotics. This may be an unrealizable pipe dream given the difficulty we have in discovering new human antibiotics.

In Our Hospitals, Things Are Getting Critical

The chart from the Centers for Disease Control (Fig. 3.2) shows the increase in multiply resistant staph (MRSA), enterococci (VRE) and *Pseudomonas aeruginosa* (FQRP) over the last 30 years in US hospital intensive care units. The ICU is the last place we want to see very resistant pathogens, and yet this is where they seem to be the most frequent.

Table 3.1 below shows the latest data from the Centers for Disease Control on frequency of resistance of these and other pathogens to key antibiotics. The data specifically represent ICU infections that were associated with devices such as intravenous catheters, breathing tubes and urinary catheters.

In Europe, the situation is not much different. The chart below (Fig. 3.3) shows the state of resistance in key pathogens isolated from bloodstream infections in Europe. The rates of MRSA infection are not very different from those in the US – hovering around 30%. About 20% of European bloodstream isolates of *Pseudomonas aeruginosa* are resistant to our last line antibiotic class, the carbapenems.

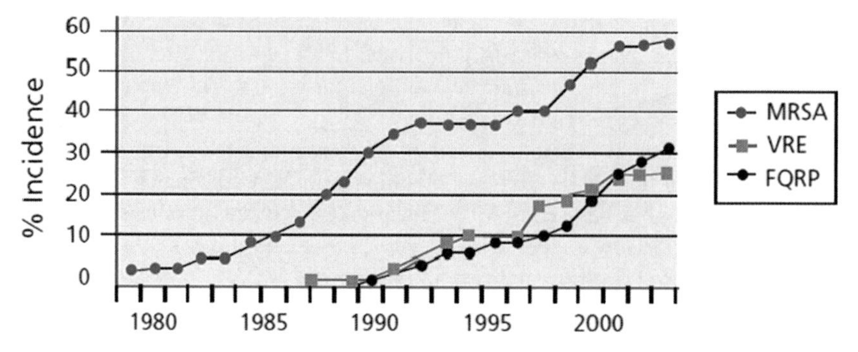

Fig. 3.2 Resistance rates in US intensive care units over time. MRSA – methicillin-resistant *Staphylococcus aureus*. VRE – Vanomycin-resistant enterococcus. FQRP – Fluoroquinolone (ciprofloxacin) resistant *Pseudomonas aeruginosa*. From the CDC and the Infectious Diseases Society of America with permission

Table 3.1 Pathogens causing device-associated infections in US hospital ICUs January 2007 to October 2007

Pathogen	%
MRSA	56
Vancomycin resistant *E. faecalis*	7
Vancomycin resistant *E. faecium*	80
Carbapenem-resistant *Pseudomonas aeruginosa*	25
Klebsiella pneumoniae resistant to new cephalosporins	25
Carbapenem-resistant *Acinetobacter baumanii*	35

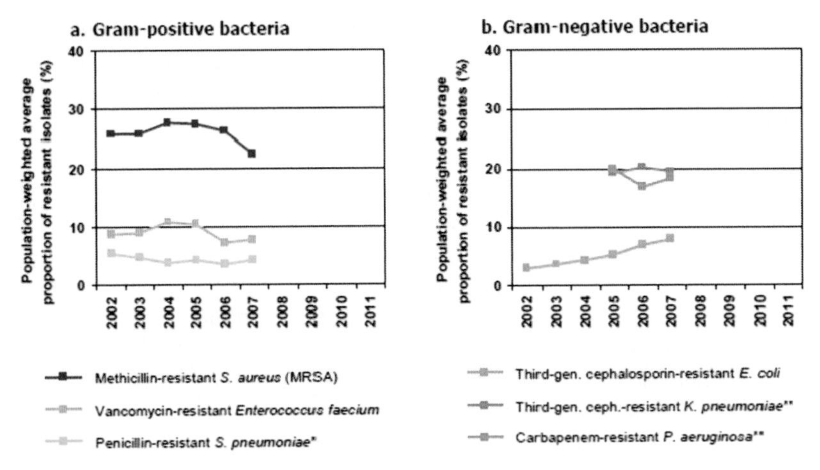

*S. pneumoniae: *excluding Greece, which did not report data on this bacterium to EARSS.*

**K. pneumoniae *and* P. aeruginosa: *excluding Belgium and Slovakia, which did not report data on these bacteria to EARSS.*

Fig. 3.3 Population-weighted, average proportion of resistant isolates among blood isolates of bacteria frequently responsible for bloodstream infections, EU Member States, Iceland and Norway, 2002–2007. Taken from The Bacterial Challenge, a Time to React, published by the European Centre for Disease Prevention and Control and the European Medicines Agency

For the MRSA and the vancomycin-resistant enterococci, we now have two antibiotics that will work. One is only available intravenously, not orally, and does not have regulatory approval specifically for treatment of enterococcal infections. Resistance to both has been reported.

The situation in Gram negative bacteria is becoming even more alarming. In years past, we had a wide choice of antibiotics active against these bacteria. The sulfa drugs and tetracycline worked. Ampicillin or the combination of amoxicillin (similar to ampicillin) plus an inhibitor of the enzyme that destroys ampicillin, B-lactamase, worked (Augmentin). Most of the cephalosporins (similar to ampicillin but with activity against a wider array of bacteria) were also effective. Hospitals had the luxury of deciding which of the many effective drugs they would put on their formularies. In many parts of the world, including the US, those days are long gone.

In many hospitals and chronic care facilities today, resistance has gotten to the point where only one (essentially) class of antibiotics is left for physicians and patients, the carbapenems. For many physicians and patients, our antibiotic of last resort has become our drug of first choice. Given the "you use it you lose it" rule of antibiotics, you can guess what is happening now. These Gram negative pathogens, especially *Klebsiella* have acquired a gene for a new enzyme (B-lactamase) that can destroy the carbapenems. Its called KPC for *Klebsiella pneumoniae* carbapenemase. The first one of these was isolated from a patient in North Carolina in 1996. The new enzyme destroys the penicillins like ampicillin, even our most modern cephalosporins, and our last line drugs, the carbapenems. KPC is not inhbited by currently marketed B-lactamase inhibitors – so those combinations like Augmentin and others are not effective. In addition, these bacteria are frequently resistant to multiple other antibiotics, even the quinolones like ciprofloxacin or levofloxacin. To treat infections by these pan-resistant strains, physicians are going back to our old friend (not), colisitin.

KPC *Klebsiella* are now spread throughout the world. We don't have good survey data for many geographic locales. (I can't understand why this is so). In New York City, about 30% of hospital *Klebseilla* carry KPC. The strains are also widespread in urban hospitals of Pennsylvania and New Jersey. Israel, Greece and China also have suffered significant epidemics of infection with these strains.

As I noted above, our antibiotic class of last resort, the carbapenems, has now become our antibiotic of first choice for many Gram negative infections. This is not a good sign. Unfortunately, about 25% of US ICU isolates of *Pseudomonas* and 35% of *Acinetobacter* are also resistant to carbapenems. In many of these cases, we are again back to colistin that I remember from my days of training in the 1970s (see Chapter 1). Because colisitin was developed and marketed so long ago, we think it may work, but we don't know how well and we know it is toxic but we don't know how toxic.

Acinetobacter has become a big problem in the military. Our soldiers wounded in Iraq and Afghanistan are being transferred back to the military hospital at Landstuhl in Germany or Walter Reed here in the US with multi-resistant *Acinetobacter* infections. One outbreak of *Acinetobacter* infection involved 70 soldiers and six medical evacuation centers and military hospitals. Ten percent of the strains were resistant to carbapenems, our last line of defense against these organisms. A separate study of *Acinetobacter* from returning soldiers showed a 36% resistance rate to the carbapenems. For most of these strains, the therapeutic choice is very limited if one exists at all. Crude mortality rates from *Acinetobacter* infections vary from 16 to 43% overall and they tend to increase when the *Acinetobacter* are resistant to multiple antibiotics.

The Infectious Diseases Society of America, an organization of physicians specializing in the treatment of infectious diseases, has developed a priority list of bacteria for which we desperately need new antibiotics and often, where we also need more information on the antibiotics we already have. Their acronym for these bacteria is ESKAPE. The list below is taken almost directly from the Society's latest publication on this topic.

> **E: *E. faecium* (VRE)** has consistently identified as the third most frequent cause of nosocomial bloodstream infection in the United States. Vancomycin resistance likewise continues to increase, with a rate of ~60% among *E. faecium* isolates.
>
> **S: *S. aureus* (MRSA).** Despite the addition of several new agents to treat MRSA infection, clinicians are routinely faced with treatment challenges involving patients with invasive disease. Although criteria for treating skin and skin-structure infection due to community associated MRSA are evolving, the need is great for oral agents for step-down therapy for the group of patients who require initial parenteral therapy. Novel classes are clearly needed for MRSA, because current drug classes exhibit treatment-limiting toxicities and emerging resistance.

K: ESBL-producing *E. coli* and *Klebsiella* species. ESBL producing strains are those that produce enzymes that inactivate most penicillin and cephalosporin antibiotics before they can kill the bacteria. Infection due to ESBL-producing *E. coli* and *Klebsiella* species continue to increase in frequency and severity. Despite this growing, serious problem, the molecules in late stage development represent only incremental advances over existing carbapenems.

More K: *K. pneumoniae* Carbapenem-Hydrolyzing Enzymes. Carbapenem-resistant Enterobacteriaceae are increasingly recognized as the cause of sporadic and outbreak infections in the United States and Europe. These organisms cause severe infections among residents of long-term-care facilities and are not easily detected in the clinical microbiology laboratory. Little is known with regard to optimal antimicrobial therapy, and few drugs demonstrate activity. Tigecycline (a relative of tetracycline active against many resistant bacteria) and the polymyxins, including colistin have been used in individual cases with variable success. There are currently no antibacterials in advanced development for these resistant pathogens.

A: *A. baumannii.* The incidence of infection due to multiply resistant Acinetobacter species continues to increase globally. Unfortunately, as in 2006, we cannot identify candidate compounds in late stage development for treatment of resistant Acinetobacter infection; this pathogen is emblematic of the mismatch between unmet medical needs and the current antimicrobial research and development pipeline.

P: *P. aeruginosa.* Rates of infection due to resistant *P. aeruginosa* continue to increase in the United States and globally, as does resistance to both the quinolones and carbapenems. Recent reports also document resistance to the polymyxins like colistin. To date, no drugs in clinical development address the issue of carbapenem resistance or offer a less toxic alternative to the polymyxins.

E: Enterobacter Species. Enterobacter species cause an increasing number of health care–associated infections and are increasingly resistant to multiple antibacterials. Other than colistin and perhaps tigecycline, few antibacterials are active against these resistant organisms, and we found no drug in late stage development for these pathogens.

Drs. Elemam, Rahimian, and Mandell of St. Vincent's Hospital in New York recently expressed their frustration in describing two cases of infection caused by *Klebsiella* resistant to all known antibiotics. They said, "It is a rarity for a physician in the developed world to have a patient die of an overwhelming infection for which there are no therapeutic options. These cases were the first instance in our clinical experience in which we had no effective treatment to offer. Trends in urban hospitals are often the harbinger of the future. We share these cases to highlight some troubling issues that soon may be relevant to increasing numbers of physicians and patients across the United States." For the ESKAPE organisms, we are going back to a pre-antibiotic era.

Antibiotic Resistance Plus Toxin Production Equals Death

The US is now experiencing a major epidemic of infections caused by a gastrointestinal pathogen called *Clostridium difficile.* This organism produces toxins that cause inflammation and fluid secretion in the lower GI tract. The disease produced ranges from mild diarrhea to severe life threatening disease. The older you are, the

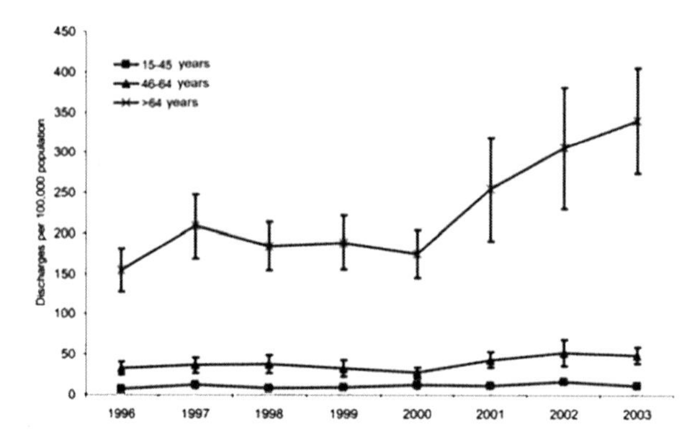

Fig. 3.4 Rates of US short-stay hospital discharges with *Clostridium difficile* listed as any diagnosis, by age. Isobars represent 95% confidence intervals. From McDonald et al., EID 2006. 12: 409

more likely you are to have severe disease. *C. diff*, as the pathogen is known in the jargon of infectious diseases specialists, is normally inhibited from growing by the normal gut flora. It most frequently causes disease when the normal gut flora is perturbed by antibiotic use. It survives because it is resistant to most antibiotics.

These data (Fig. 3.4) from the CDC show that the rates of *C. diff* diarrhea have increased about 100% between 1996 and 2003 and that virtually all of this increase occurred in patients older than 64 years of age. We find the same picture when we look at deaths from *C. diff*. The death rate has climbed tremendously over the same time period and it is mainly coming from the older population. In 1993, *C. diff* was associated with an 8% chance of death while in 2003, the death rate was between 9.5 and 10%. Most of this mortality was among those over 64 years of age where the death rate in recent years ranges from 30 to 50%. It is estimated that the cost for *C. diff* alone in the US is over $1 billion per year and climbing rapidly.

Of course, the CDC is giving us only a picture of what is happening in acute care hospitals. There is also disease occurring in long-term care facilities and in our communities.

Why have we had this sudden increase in cases and in severity of *C. diff* disease? Part of the explanation is that much of this increase is related to one or two strains now spreading worldwide that seem to be better able to disseminate and are more virulent. This seems also to be a disease of antibiotic resistance coupled with the other normal effect of antibiotics – a perturbation in the normal flora of the gut.

All of these problems are interrelated. For example, the treatment of choice for serious *C. diff* disease is vancomycin administered orally. Vancomycin is not absorbed from the gut – so its concentrations there are very high. These high vancomycin concentrations may tend to select for enterococci, a normal gut inhabitant, resistant to vancomycin. Such strains are often multiply antibiotic resistant and can also cause infection.

Some believe that with good hygiene, handwashing, meticulous cleaning, somehow all these infections in hospitals will disappear or at least those caused by resistant pathogens will somehow diminish. While we might be able to reduce the incidence of these infections in hospitals, we will never be entirely rid of them. For example, even in the best hospitals under the best of circumstances, the highest rates of handwashing between patients are only around 70–80% with most hospitals averaging around 30–50%. For many reasons, perfection is unlikely to be achieved. Also, there is no real evidence that many of the infections that occur in hospitals are actually preventable either by good hygiene or by improved environmental cleaning procedures. Therefore, although it is easy to blame hospitals and their staffs for infections that are acquired there, no matter how good our infection control procedures become in the hospital, we will always need to have new antibiotics active against resistant strains. The corollary to this is that punishing hospitals for hospital-acquired infection financially, as currently proposed for medicare, will have little impact on true infection rates. All it may do is encourage hospitals to hide the truth.

The Infectious Diseases Society is worried. They have even formed a Task Force to deal with the conundrum of rapidly growing antibiotic resistance and the lack of new products available to treat these infections. We will discuss this in more detail in a later chapter.

Our Communities are Not Spared

We all agree that hospitals are to be avoided if possible. Are we safe from antibiotic resistance in the community? Most of the antibiotics that we use every year for human health are used in our communities. The principle of "you use it, you lose it" therefore would predict that our communities would not be spared either.

Let's look at a common cause of urinary tract infection (UTI) in our communities, *E. coli*. This is a normal inhabitant of our intestinal tract that occasionally gets somewhere it shouldn't be. If it's equipped with the right stuff (virulence factors), it can cause disease. Antibiotics that have been commonly used to treat UTI include ampicillin, bactrim, and ciprofloxacin. Table 3.2 below shows resistance rates to those antibiotics.

The isolates from the US tended to be more resistant than those from Canada. The resistance to ciprofloxacin, a quinolone, is increasing with time, especially among

Table 3.2 Antibiotic Resistance in outpatient urinary isolates

North American UTI Collaborative Alliance 2005	
E. coli	% Resistant
Ampicillin	38
Bactrim	21
Cipro	5.5

urinary tract isolates. Most clinicians believe that these resistance rates would preclude routine therapy with ampicillin and bactrim. While ciprofloxacin remains a viable option for therapy in the US, resistance is increasing and there are occasional isolates where even ciprofloxacin would not work.

Essentially all the antibiotics used for UTI in communities are under threat with the possible exception of nitrofurantoin. Nitrofurantoin is an old antibiotic and we still do not have a good understanding of how it works. Resistance remains rare. The problem with nitrofurantoin is that, although it works well against *E. coli*, it does not work reliably against many of the other bacteria that cause UTI. Nitrofurantoin may also be more toxic than the other drugs. For these reasons, its use diminished with the availability of ampicillin, bactrim and the quinolones. Now that *E. coli* is becoming resistant to these drugs, we still have nitrofurantoin in our back pocket. Unfortunately, using this drug would require physicians to obtain a sample of urine for culture to be sure that the bacterial cause is in fact *E. coli* or another susceptible organism. Physicians don't often do that anymore.

These data are all consistent with our hypothesis – antibiotic use begets resistance. Animal use of fluoroquinolones probably does not help.

Strains of staph resistant to our most powerful antibiotics have now also invaded our communities. *Staphylococcus aureus* has an interesting history. In the pre-antibiotic era, severe infections with staph like pneumonia and bloodstream infections were almost invariably fatal. When the sulfonamides and later penicillin came along, we were suddenly able to cure most of even the most severe infections. In the case of penicillin in particular, though, resistance was recognized even before the antibiotic was introduced for military use. As I mentioned in a previous chapter, penicillin-resistant staph started out causing infections in hospitals. With the pressure of widespread penicillin use and the easy transmission of staph from hospital patients and employees to family and friends in the community, penicillin resistance quickly spread. By the 1970s, most strains of staph were penicillin resistant whether or not they were causing infections in the hospital or in our communities. Luckily, right about this time, several antibiotics appeared on the market that worked against the penicillin resistant strains. They included a new penicillin derivative called methicillin, the cephalosporins which are related to the penicillins and a drug called vancomycin. Shortly after the introduction of methicillin, the first strains of methicillin resistant staphylococci, called MRSA for methicillin-resistant *Staphylococcus aureus*, were described in Great Britain. One of the problems with MRSA is that they are frequently resistant to most antibiotics used to treat staph infections including the cephalosporins, tetracyclines, etc. They remain susceptible to vancomycin. Vancomycin is an antibiotic that can only be administered intravenously.

In the 1980s, the US began to experience an epidemic of MRSA infections in hospitals. Figure 3.2 shows that in the early 1980s, only about 10% of ICU isolates of staph were MRSA. Today, that number is over 50% and the MRSA is not just limited to the ICU. MRSA is found in all areas of our hospitals. Vancomycin has become the drug of choice for the treatment of MRSA infections. Linezolid was introduced to the market in 2000 and in 2003 daptomycin was marketed. Both are also active against MRSA but only linezolid can be given as a pill.

Just as was seen for penicillin resistance in staph, MRSA have now spread to our communities. In a recent CDC study examining staph isolates from emergency rooms across the US, over 60% of strains were MRSA. Most of the isolates in the US actually represent a single strain called USA 300. It is not yet clear what it is about this strain that allows it to carry resistance and cause infection in the community. The story though is again familiar. When penicillin resistance spread to the community, it too was frequently carried by a single strain of staph that seemed well adapted to spread and which was very virulent. That particular strain disappeared after about 10 years for reasons we still don't understand. The penicillin resistance in the community remains, however, probably now carried by multiple different strains. Perhaps we will see the same thing with USA300.

There is one interesting difference between hospital and community MRSA. Although hospital strains of MRSA tend to be resistant to almost all other antibiotics, strains originating in the community tend to remain susceptible to bactrim, tetracycline, clindamycin and certain other antibiotics. None of these antibiotics has been carefully studied for the treatment of staph infections. But because they are inexepensive generics and can be taken orally, physicians often prescribe them for community acquired MRSA infections. There are ongoing trials of these cheaper generic agents in the treatment of MRSA infections, but we don't yet have the results.

Another interesting aspect of the community acquired MRSA is that they are starting to show up in hospitals. Many of the isolates appearing in hospital microbiology laboratories are not the usually multiply resistant variety, but the more susceptible community strains of MRSA. Whether this will allow us to modify the way we treat patients or not remains to be seen.

Resistance – Summing Up

If we can manage to bring new antibiotics to the marketplace, should we anticipate resistance to them as well? The answer is a qualified yes. Although bacteria eventually seem to be able to become resistant to whatever antibiotic we use, the time it takes for this to occur varies but can be very long. For example, resistance to the penicillins was discovered before the drug was ever marketed. Vancomycin, on the other hand, was marketed in 1956 and resistance was not seen until 1986. Another issue is that resistance may arise in one species but not in another. Again, vancomycin resistance is extremely common in *Enterococcus faecium*, but is much less common in its very close cousin *Enterococcus faecalis* even though both organisms live side by side in the gut and can easily exchange genes. We don't know why. In fact, since the discovery of vancomycin resistance in enterococci in 1986, we have all been waiting for its appearance in staph. Scientists have carried out experiments showing that it is easy to transfer the resistance from enterococci to staph in the laboratory, even on the skin of animals. Although there have been a few cases of human infection with vancomycin-resistant staph, these remain exceedingly rare.

Again, we don't know exactly why. When a new antibiotic is marketed, it is often very difficult to predict how long it might take for resistance to appear. If only every antibiotic could last as long as vancomycin!

One of the questions that scientists and economists have struggled with over the years is trying to put a specific price tag on resistance. Most studies show that morbidity like length of hospital stay and mortality are increased with antibiotic resistant infections compared to infections in similar patients with susceptible strains to about 1.5–2-fold. One recent study from South Carolina showed that for the authors' hospital, there was a 30% increase in costs for such patients. Of course, these calculations do not take into account years of life lost or quality years lost, which would add enormously to the overall costs to the US. When you consider that 2 million Americans acquire infections in hospitals every year and that anywhere from 30 to 70% involve resistant bacteria, you can see that the costs could quickly skyrocket. Europe, as noted in a later chapter has also tried to tackle this question and they come up with costs in the billions of dollars.

Antibiotic resistance is a societal problem. It derives from the way we use antibiotics, the way we dispose of our garbage and our sewage and our hygienic practices at home, at work, in schools, in long term care facilities and in hospitals. Resistance costs us lives of loved ones, lost productivity and real dollars in terms of the increased care required for these patients. We can only truly address the problem as a society. Even if we improve our approach to antibiotic use and improve our hygienic practices, it is unlikely that we will be able to solve the problem of the continued selection of antibiotic resistance. We just do not have enough information to know how to halt this natural progression entirely. Therefore, as it stands today, antibiotic resistance also requires that we constantly have new, effective antibiotics coming on line in the marketplace so we can treat patients with resistant bacterial infections. No one ever wants to be in the place of Drs. Elemam, Rahimian, and Mandell or the patients they described. Only we as a society can create conditions where we can have such a pipeline of new antibiotics.

Chapter 4
The FDA

We Need Them, but They Have Become Part of the Problem

It is clear that we need a viable, strong, active and even interventionist FDA. But we also need new antibiotics. How did we get to a place where these two obvious needs might be in conflict?

The Food and Drug Administration had its origins in tainted food and agricultural products, contaminated antisera and bogus medicines sold to the public and to the US armed forces in the nineteenth century. The Bureau of Chemistry in the Department of Agriculture hired a chemist, Harvey Washington Wiley, in 1883. He worked to bring adulterated food products to public attention and then to study the effects of the adulterants in human subjects using his "poison squad" of volunteers. These revelations led in large part to the passage of the Food and Drugs act of 1906 – it was called the Wiley Act at the time. The law was mainly directed at appropriate labeling and making sure that additives and compounds were adequately pure and well described. The Bureau of Chemistry was charged with its enforcement.

After the election of FDR in 1932, it was becoming painfully clear to the public and to government that the 1906 law needed updating. With only labeling of ingredients as its mandate, the agency could not remove toxic or ineffective products from the marketplace. Several scandals brought this major shortcoming to public attention. An eyelash enhancer was causing severe reactions and even blindness. A worthless "cure" for diabetes was being sold. Finally, Elixir sulfanilamide was promoted for use in children. It had the sulfa antibiotic all right, but it was dissolved in a sweet tasting antifreeze derivative that killed 100 people, many of them children. The Food, Drug and Cosmetic Act of 1938 addressed these issues by requiring that drugs be approved before marketing. The agency was given powers to enforce prohibition of false claims and tolerance limits for certain noxious substances were mandated. Prescriptions were required for many drugs, including the new sulfa antibiotics.

The Pencillin Amendment was passed in 1945 in response to a large number of penicillin analogues being introduced to the market. The law required testing of safety and efficacy of all penicillin analogs, and ultimately all other antibiotics. The amendment was rescinded in 1983 as it was then considered superfluous.

D.M. Shlaes, *Antibiotics*, DOI 10.1007/978-90-481-9057-7_4,

As a result of the thalidomide tragedy, where the drug was used for the treatment of nausea in pregnancy and resulted in untold numbers of birth defects, the Kefauver-Harris Amendment was passed in 1962. It required drug manufacturers to prove efficacy as well as safety during pregnancy. This law greatly expanded the powers of the FDA.

A timeline of key events in the history of the FDA focusing on antibiotics is shown below (Fig. 4.1).

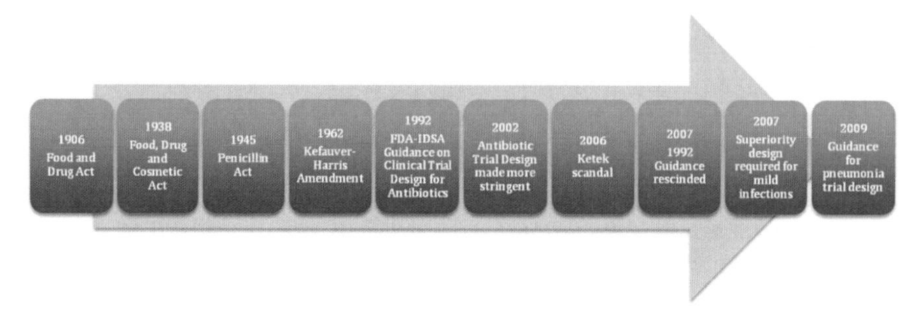

Fig. 4.1 A timeline of key developments at the FDA emphasizing those affecting antibiotics

I don't think that anyone can question the necessity of the FDA and the enormous good it has done for patients and physicians since it's founding. For antibiotics, though, things are starting to unravel.

Sulfonamides and penicillin were tested in clinical trials on only a few hundred patients with a variety of infections and were used topically, orally and intravenously. In some trials, untreated patients with similar infections seen at the same time as treated patients served as controls. In other trials, the only controls were untreated patients seen in the past with similar infections and where the same sorts of data were collected. Such patients would be historical controls. Although the FDA still accepts historical controls for some trials, especially in rare diseases or where the historical database is exceptionally large and robust, they are almost never used anymore to study antibiotics – and rightly so. We were so lucky. The marketed compounds went on to be used in the treatment of untold millions of patients over the years. No one doubts the utility and importance of these early antibacterial compounds for human use (except, perhaps, the FDA itself).

These days, drugs are withdrawn for serious side effects occurring at rates of 1 in 100,000 treated patients. Our clinical trials for antibiotics require us to treat just a few thousand patients before the drug is approved. Therefore, there is no way we would be able to detect rare but potentially serious side effects. Imagine what kind of a chance we were taking when drugs were approved after use in only a few hundred patients.

During the golden age of antibiotics between 1955 and 1985, as described in Chapter 6, Industry, FDA and the Infectious Diseases Society worked closely together to define how trials should be designed and evaluated. Recently, things have gone less well. The figure below shows approval of new antibiotics between

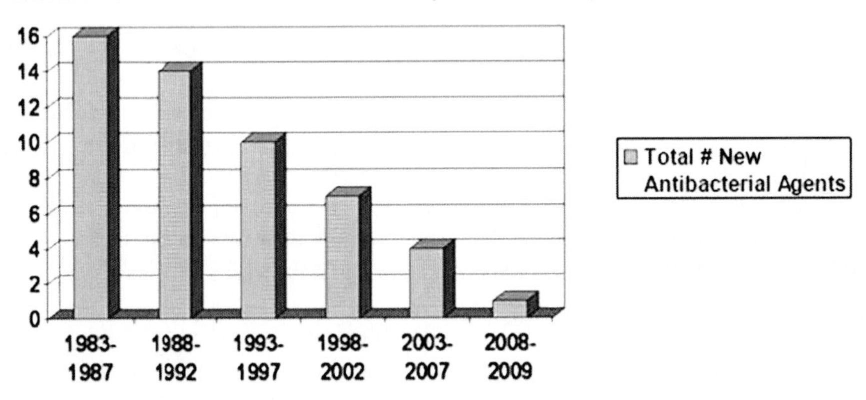

Fig. 4.2 Approval of antibiotics at the FDA over time. From the Infectious Diseases Society of America with permission

1983 and 2007 (Fig. 4.2). 2008–2009 is looking to be even worse. The Infectious Diseases Society of America, in their white paper, Bad Bugs, No Drugs, points out that between 1998 and 2004, only 10 new antibiotics were approved. In 2002, 89 new medicines were approved, but none of them were antibiotics. Since 2002, only eight new antibiotics have been approved with no approvals occurring in 2006 and again in 2008. This adds up to about one new antibiotic approved per year for the last 7 years compared to over three antibiotics per year between 1983 and 1987. Of six antibiotics submitted to the FDA between 2007 and 2009, only two have been approved. This represents a historically high late stage failure rate of 67% for antibiotics.

Does the FDA contribute to our lack of new antibiotics? In my opinion, the answer to that question is a resounding YES. Without significant changes from the FDA and perhaps from Congress, the lack of new antibiotics can only be expected to worsen. In 2002, Bob Moellering and I wrote an article in the journal Clinical Infectious Diseases entitled The FDA and the End of Antibiotics. In it, we expressed the concern that the uncertainty around requirements for clinical trial design, the tightening of trial requirements with its associated increased costs and the perceived hostility of the agency towards antibiotics would help drive more companies out of the area. Since then, unfortunately, our predictions have all been correct. Our antibiotic pipeline has only become drier and the number of companies active in antibiotics research has continued to dwindle.

How does the FDA contribute to the perfect storm for antibiotics? Guidance documents from the FDA can be helpful in that they tell everyone the kinds of things they need to do in order that the FDA will approve the product in question. They are currently issuing a series of guidances for the design of clinical trials for antibiotics that make it difficult, and at times impossible, to actually carry out the proposed trials. But if these guidance documents mandate studies that are not feasible – where are we? Nowhere.

The FDA seems to be buffeted by political pressures that are frequently based more on rumor and innuendo than good science. They allow themselves the luxury of changing policy while sponsors are in the midst of multi-10s of millions dollar trials invalidating the very designs they agreed to at the start of those trials. They are inconsistent in their treatment of generic compared to branded antibiotics where the latter get much more scrutiny. The FDA does not always avail itself of the best advice even though, given their recent leadership, they need the best advice they can get.

One of the great things about antibiotics is that the tests we use both in the test tube and in experimental animals predict with a great degree of certainty whether the antibiotic will work in people or not. Once we know just a little bit about how the drug behaves in people, whether and how it is absorbed and distributed to human tissues, we can quickly, based on test tube and animal work, predict the correct dose to treat human infections. The FDA is very well aware of this. Included in the documents companies must submit to apply for permission to study the drug in patients, the FDA requires an entire section on how the company chose the dose they propose to study. This should remove a tremendous worry from companies, regulators, physicians and patients in the sense that we know that the antibiotic will work and we have a good idea of the dose that will be required in order for that to happen. Of course, we can't always predict human toxicity or metabolism with this degree of accuracy, so antibiotics can fail because of problems in those areas. They can also fail even if they work at the doses chosen, but they don't work as well as the antibiotics to which they are being compared in the clinical trials.

The FDA Increases Clinical Trial Design Stringency and Costs. Companies Abandon Antibiotic Research

In 1999, I was working at Wyeth Pharmaceuticals. Wyeth had discovered a new chemical series of tetracycline derivatives that was active against a variety of resistant bacteria including those resistant to the tetracyclines. We were getting ready to enter the last (and most expensive) stage of clinical trials prior to submitting our request for marketing approval. We followed the FDA guidelines in designing our trials. They allowed us to set the statistical stringency of the study, within limits, according to the size of the trial. Therefore, as the trial size got smaller, the stringency would be lower. In our discussions with the FDA, it became clear that they were going to require trials of a size and cost that was unprecedented for antibiotics being developed to treat patients in hospitals. The FDA was concerned that the kind of clinical trials used for approval of antibiotics in the past were not sufficiently robust. In these trials, since using a placebo is generally thought to be unethical, we compare the new, experimental antibiotic with an older, proven antibiotic. But, since the older, comparator antibiotics (after the sulfonamides and penicillin) had never been compared to placebo (no active therapy) the FDA felt that there was a chance that new products might slip to a point of being ineffective. The issue is a statistical one (but has little to do with life as we know it!). When you compare two products in a trial, the two are compared statistically. This being the case – you can't really do an equivalence trial since to prove that two drugs are statistically equal requires an infinitely high number of patients. For antibiotics, you also would be hard pressed to do a superiority trial since the antibiotics work so well that proving superiority also requires a very high number of patients. It would also be hard to recruit patients into a trial where you know that the patient has a pathogen resistant to say the comparator antibiotic but not the new one given that the patient would have a random chance of receiving either drug. Therefore, all antibiotics since the 1950s have been studied in a so-called non-inferiority trial. In this sort of trial, you must accept some limit on the chance that the new drug is inferior to the standard approved therapy used as a comparator. That statistical margin is the problem. The tighter the margin, the more patients are required for the trial. In the past, as I noted above, margins were deliberately set to keep trials at a reasonable size. These margins, usually set somewhere between 10 and 20%, were acceptable to the FDA. 15% was a commonly used margin. That is, for seriously ill patients, the FDA would accept the possibility that the new treatment might be up 15–20% worse than the comparator. So, over time, in the worst case scenario for every drug, if one were 15% worse than the other and than that one became the comparator for the next and so on, the new antibiotics could end up being no better than nothing at all. Those of us actually treating patients with infections were always puzzled by this worry since we knew from the evidence of our own eyes that the antibiotics were effective for most serious infections. The FDA wanted us to reduce that margin to 10%. When you compare the patient numbers for the two circumstances, the 10% margin required more than twice the patients (and therefore about twice the cost) compared to the 15% margin. Remember, we are talking about a theoretic maximum difference between the two drugs, not the actual difference. So, while scientifically, it might be preferable to narrow this margin (why not to 5 or 2 or 1%?), the trials were going to cost more, expose more patients to an experimental drug and were going to take longer to perform. The time until the new therapy would be available to patients would be longer and the time to market for the new therapy would be more distant.

We presented the result of our discussions with FDA to Wyeth's senior management. When they saw the estimated costs for the FDA-proposed trials and the increase in time required, they balked. Our management was concerned that it would take too long for us to recoup the trial costs or that we might never recoup them. They put the entire program on hold.

We worked with Pharmaceutical Research and Manufacturers of America (PhRMA, the pharmaceutical trade and lobbying group) and with the Infectious Diseases Society of America representing infectious diseases physicians to open a public discussion with the FDA in hopes of salvaging our potentially important new antibiotic and to keep open opportunities for the development of other new antibiotics. A series of workshops with all three, FDA, PhRMA and the Infectious Diseases Society were held during 2000 and 2002 to address FDA concerns. The result of these discussions was that the FDA did not make any blanket decision covering all antibiotics, but agreed to evaluate trial designs on a case-by-case basis.

In further discussions with Wyeth the FDA allowed Wyeth to proceed with 15% margins and tigecycline was developed and finally approved in 2005. Wyeth as it turns out, was one of the last companies to be allowed to use a 15% margin – almost everyone else since then has had to use a 10% margin for most indications.

Between 2000 and 2002, Roche, Lilly, Bristol-Myers Squibb and even Wyeth all announced they would discontinue research in antibiotics. Many more would follow closely on their heals. There are many reasons for this as we will discuss in Chapter 6, but one of the reasons certainly was the uncertainty around the FDA's clinical trial requirements and a feeling among many in industry that the agency was actually hostile to antibiotics in general.

One of the requests industry and the Infectious Diseases Society made to FDA in 2000 was that the agency modernize their guidance for the development of antibiotics so at least companies would know what kind of trials they would have to conduct to obtain approval. Since then, the agency has actually released a number of new draft guidance documents. The good news is that industry knows what it has to do. The bad news is that, frequently, they can't do it.

The first new guidance released indicated that all the old guidance documents on trial design for antibiotics were no longer considered valid by the agency. Next, the agency required a justification for the statistical margin that was to be used in proposed comparative trials. I think this was a way of getting sponsors to help FDA in doing some literature research in this area. The idea is that to define this margin, you have to define the benefit that the antibiotic would have compared to placebo. Since placebo controlled trials have not been done since the sulfonamides and penicillin, this gets to be a bit difficult.

With their advisory committee and in public, the FDA began to examine the issue of antibiotics used for mild infections like sinusitis, bronchitis and otitis (middle ear infections). The issue for these infections is that they frequently are caused by viruses and not bacteria and therefore would not respond to antibiotics in any case. This leads to much of the unnecessary use of antibiotics which in turn probably leads to antibiotic resistance. The other question is that even when bacteria cause these types of infections, will they get better without treatment? Will serious complications arise without antibiotic treatment? How do we know that antibiotics even work? The scientific literature is very conflicted on this subject. The area of mild infections is directly related to the agency's basic concern about comparative trials where a placebo is not used. How do we know that the standard or comparator antibiotic is better than no antibiotic?

"Mild" Infections Require Placebo-Controlled Trials – Industry Balks

Otitis media or middle ear infection might be the clearest example. These are the typical ear infections occurring mainly in childhood starting at around 6 months of age. Otitis media is painful and for many years clinical practice in the US was to treat them with antibiotics in the belief that killing the bacteria that cause the infection would result in more rapid relief of pain and would prevent potentially serious complications. Many parents in the US have had the experience of taking their sick child to their physician or to an emergency room for these infections. Some children who had repeated episodes were even given antibiotic prescriptions in a "just in case" sort of arrangement. If they had typical symptoms, they would call their doctor and start antibiotics until they could get into the office. It goes without saying that many antibiotics have FDA approval for their use in otitis media – all based on trials comparing one antibiotic with another and none with a placebo control. For the pharmaceutical industry, otitis was a very lucrative market.

However, a number of clinical trials comparing antibiotic to placebo were carried out, mainly outside the US, which seemed unable to show a clear advantage of antibiotic therapy over a simple prescription of a pain reliever. In some of these trials, it appeared that infections caused by one particular organism, *Streptococcus pneumoniae*, required antibiotic therapy for cure. But this organism only caused a minority of all otitis and there is now a very effective vaccine that protects, to a certain extent, against otitis caused by *S. pneumoniae*. Many studies later, it seems that the best approach is one of expectant therapy. The child is given a pain-reliever. If they still have symptoms after 2–3 days, an antibiotic is prescribed. In these circumstances, around 80% will not need antibiotics. There is no difference in any outcome between patients given antibiotics immediately and those treated expectantly. The agency now requires a placebo-controlled trial to prove that an antibiotic works in otitis. Given the data, we can agree that this is a reasonable requirement.

Some pediatric infectious diseases specialists disagree that patients with true otitis media do not need antibiotics. They argue that the diagnosis in many of the placebo-controlled trials that were carried out in the past were faulty and did not represent true bacterial infection of the middle ear. They claim that antibiotics can play an important role in shortening the duration of disease and preventing complications in true otitis media. This disagreement has led to an ongoing placebo controlled trial in Finland that is being funded by Finland. The patients are very carefully examined such that the diagnosis of otitis is not in doubt. This may be the first trial where there will be no argument about whether the patients actually have otitis and where there is an untreated control group. If this trial shows a significant benefit for antibiotics, it is possible that placebo controlled trials will be a thing of the past. If there is no benefit, of course, it will mean that for most patients antibiotics are not necessary. Whether such data would alter clinical practice and patient attitudes is another question.

The American Society of Pediatrics tried to promulgate guidelines suggesting that patients with severe symptoms, those age <6 months and those where the diagnosis is certain that it is otitis media be treated with antibiotics immediately. This leaves older children, those with milder disease and those where the diagnosis is less certain (the majority of patients) available for expectant therapy. In spite of these guidelines, recent surveys have shown that only 15% of children in the US are treated expectantly. That number should be about 85%. The most common reason is parental concern (85% of parents) about not using antibiotics. In addition, physicians prefer to treat based on probability of infection as opposed to certainty. One approach some physicians have been taking is to give the parents a prescription to fill in case the expectant therapy – pain reliever – isn't enough. That way, the parent controls the destiny of their child and themselves, the doctor has provided specific therapy, and everyone is satisfied.

The major need for new antibiotics in pediatrics today might be for those children allergic to the penicillins and their relatives. Many of the pathogens that cause otitis are resistant to the other types of antibiotics approved for use in children with otitis and clinical failures do occur because of these resistant organisms. I asked a highly respected colleague working in pediatric infectious diseases how he handles this dilemma. Simple, he replied, I just use a quinolone antibiotic. He believes at least one of them is safe for use in children even though they have never been approved for treating otitis in children and other pediatricians and the FDA have expressed safety concerns about the use of quinolones in children.

Given this state of affairs, I doubt that industry will attempt to develop new antibiotics for otitis in the foreseeable future. Of course the market loss for industry is a large one. But the reduced pressure selecting for resistance by unnecessary use in otitis is a benefit for new antibiotics developed for other kinds of infections.

Sinusitis is also a large potential market for antibiotics and is more controversial. As is the case for otitis, many antibiotics are already approved by the FDA and marketed for the treatment of sinusitis based on comparative trials. According to the American College of Physicians, in most cases, antibiotics should be used only for patients with the specific findings of persistent purulent nasal discharge and facial pain or tenderness who are not improving after 7 days or those with severe symptoms regardless of duration. This recommendation is based on a number of placebo-controlled clinical trials where a modest benefit from therapy either in terms of cure or in decreasing length of illness was mostly offset by an increase in adverse effects by the antibiotics when compared to placebo. The FDA has responded to this by saying that, given the modest treatment effect, they would be unable to judge, statistically, whether a given antibiotic was inferior or not to placebo in the absence of a placebo control. However, since not treating patients with severe symptoms or symptoms lasting more than 7 days goes against medical guidance, it is difficult if not impossible to carry out the placebo controlled trials mandated by the FDA. The industry is staying away from this one. If we wanted a new antibiotic now or in the foreseeable future for the few cases of acute bacterial sinusitis that might be caused by resistant strains of bacteria, we would be disappointed. Again, like otitis, this was previously a large market segment for the pharmaceutical industry that has now virtually disappeared for new products.

Finally, there is bronchitis. This is a really controversial area. Patients with chronic lung disease, specifically, chronic obstructive pulmonary disease or COPD, have ongoing breathing problems and other symptoms like productive cough that get worse (exacerbations) from time to time. They are chronically colonized with bacteria in many cases. That is, even when they are not experiencing worsening symptoms, they have bacteria living in their lungs. Their exacerbations seem to be associated with the acquisition of new strains of bacteria in their lungs. For many years, physicians have thought that treating the bacteria isolated from the sputum (bronchial and lung secretions these patients cough up) of patients at the time of an exacerbation would shorten the duration of the episode and help avoid more serious complications like respiratory failure and pneumonia. Like otitis and sinusitis, many antibiotics marketed today are indicated for the treatment of these exacerbations all based on comparative trials without placebo controls. We now know that for so-called mild to moderate exacerbations, antibiotics appear to offer little advantage compared to no antibiotics. However, for more severely ill patients, studies suggest that antibiotics have an important role in reducing relapses, complications and in reducing mortality. According to the Cochrane Review, an analysis of many placebo-controlled trials for this disease showed a clear benefit for antibiotic treatment.

> For COPD exacerbations with increased cough and sputum purulence antibiotics, regardless of choice, reduce the risk of short-term mortality by 77%, decrease the risk of treatment failure by 53% and the risk of sputum purulence by 44%; with a small increase in the risk of diarrhoea. .... this review supports antibiotics for patients with COPD exacerbations with increased cough and sputum purulence who are moderately or severely ill.

Again, the FDA, in spite of this sort of information, has required placebo-controlled trials for new products. Since it might not be ethical to withhold antibiotics from patients with severe exacerbations and since antibiotics might not work as well for mild disease, only one pharmaceutical company has yet ventured into this area. Their trial did show a benefit for their antibiotic, but failed to recruit a sufficient number of patients to satisfy the FDA. They went out of business shortly after their trial was stopped. (see Chapter 6 for more details). For patients and their physicians, this will mean no more new antibiotics for bronchitis. For the industry, another market has been closed. There are academic investigators attempting to conduct a placebo-controlled trial for bronchitis with funding from the NIH, but it is not clear how severely ill the patients are who are included in the trial. No data has, as far as I know, yet been published on the ongoing NIH sponsored trial.

Here are three indications where physicians perceived that there was a routine requirement for antibiotics and where that perception has been called into question. It also seems clear that, at least in some circumstances, for sinusitis and bronchitis, antibiotics are useful and in the latter case may even save lives. If we as a society agree (I do) that our current antibiotic armamentarium is sufficiently robust and that bacterial resistance is not a problem in these diseases, then I guess there is no need to further question the FDA's current stance. Unfortunately, since, in order to develop new drugs, the industry has to think 7–15 years ahead, if antibiotic resistance were to arise as a problem for, say bronchitis, we would be without important new therapeutic options for years to come.

New Antibiotics for Mild Infections Are Forced from the Market While Generic Antibiotics Are Still Approved in the Absence of Placebo-Controlled Trials

The Ketek Scandal

As I noted earlier, there are lots of antibiotics, including penicillin, approved for otitis, sinusitis and bronchitis based on the old approach (comparative rather than placebo-controlled trials). Some of these older antibiotics even have some level of toxicity. According to the FDA's own calculus, these products have a risk:benefit ratio of zero since their benefit has never been shown using superiority or placebo-controlled trials. Has the FDA moved to remove marketing approval for these indications from these older antibiotics? No. This point was driven home recently by the scandal over the FDA handling of a new antibiotic, Ketek (telithromycin). Even Congress got involved. This is a story I have followed closely and I was present at the final FDA meeting dealing with this new antibiotic. The Ketek story illustrates the effect of political pressure on the FDA process, FDA's inconsistent treatment of branded compared to generic antibiotics, and, in my view, their lack of leadership in general.

Ketek, or telithromycin, is an antibiotic designed to overcome antibiotic resistance in respiratory pathogens. It provides an important alternative to therapy for patients who are allergic to the penicillin type antibiotics or who cannot tolerate the quinolone antibiotics. In 2004, Ketek was approved by the FDA for use in community-acquired pneumonia, acute bacterial sinusitis and in acute bacterial exacerbations of chronic bronchitis. At the public meeting to discuss Ketek, held in Silver Spring, Maryland December 14–15, 2006, the FDA positioned Ketek as an antibiotic of questionable efficacy for the treatment of acute bacterial sinusitis and acute bacterial exacerbations of chronic bronchitis, since approval was granted based on non-inferiority rather than placebo-controlled trials. They described rare but serious side effects including cases of severe liver toxicity attributed to Ketek. In this context, the FDA asked their advisory committee to weigh the risk to benefit ratio of Ketek in sinusitis and bronchitis.

Ketek is an antibiotic distantly related to erythromycin. It has the advantage of being active against erythromycin-resistant strains of bacteria.

Erythromycin-resistance (also resistance to azithromycin (Zithromax) and clarithromycin (Biaxin)) is a big problem among bacteria that cause respiratory infections like otitis, sinusitis, bronchitis and pneumonia. It has been shown that such bacteria do not respond well to therapy with the usual macrolide antibiotics like zithromax and biaxin. Ketek would then be a good choice for such patients where they might have a penicillin allergy or they might be unable to take the other major class of antibiotics for these infections, the quinolones like levofloxacin (Levoflox).

Ketek was approved in 2004 after a long regulatory history where several toxicity signals were seen in the late stage (Phase III) trials. However, post-marketing surveillance, primarily in Europe, but also in other ex-US countries where Ketek had already been sold for a number of years, showed no substantial safety problems during almost 4 million courses of therapy. Thus the agency approved Ketek for treatment of community-acquired pneumonia, sinusitis and bronchitis.

This was a scandal-ridden approval. In their analysis of the data for Ketek in 2001, the FDA requested additional safety data from the sponsor (Aventis at that time). The FDA was particularly concerned about possible liver toxicity, cardiac toxicity and visual effects that might be associated with Ketek. Aventis then carried out a 24,000 patient safety trial of Ketek. To my knowledge, this remains the largest such trial ever performed by the industry and it was performed in record time – about 1 year. Of course, the size and speed of the trial must have stretched Aventis' resources to the breaking point. This trial was so tainted by fraud among clinical investigators, one of whom was convicted and imprisoned, and by other issues with the data per se, that the FDA declared it would be unable to use any of the data for approval. This was clearly a trial too big for its britches.

It was then that the FDA turned to the voluntary safety reporting system maintained by countries where Ketek had been approved and where the drug was already marketed. This was clearly a deviation from standard FDA practice. Critics charged that since the database used was a voluntary one and was known to underestimate toxicity, it could not be relied upon for approval. A key FDA safety officer declared at the 2006 meeting that accepting the data collected by Europe was tantamount to accepting data collected by third world countries. An FDA medical officer became a whistle-blower leaking documents and internal e-mails. He accused his supervisor of inappropriately using the controversial 24,000 patient safety study in consideration of approval and of inappropriately pressuring the staff writing reports and opinions during the approval process. He further noted that the FDA knowingly failed to disclose key issues of fraudulent data to their advisory committee in 2004. The FDA supervisors responded that they were unable to do so since there was an ongoing investigation into fraud and that the enforcement arm of the FDA restricted them to silence on this issue.

Of interest, a representative of the European regulatory agency (EMEA) was present at the 2006 FDA meeting and presented data from Europe and recent European decisions related to Ketek. In Europe, approved drugs are routinely re-examined on a regular basis. (We will come back to this in our chapter on Modest

Proposals, Chapter 7). The European Medicines Agency (EMEA – the FDA for Europe) had just completed its review of Ketek including substantially more safety data than was available at the time of the 2004 approval of Ketek by the FDA. Based on its review, the EMEA approved continued marketing of Ketek for all its indications for an additional 5 years.

Since the approval of Ketek in 2004, as noted earlier, the FDA has concluded that comparative trials (used for approval of every antibiotic in history for every bacterial respiratory infection studied to date) are not sufficient to prove efficacy for otitis, sinusitis and bronchitis. Placebo controlled trials are now required. Thus, since Ketek had not demonstrated efficacy in sinusitis or bronchitis in this way, the FDA asked advisors to help determine if the risk:benefit ratio for Ketek justified continued marketing for its approved indications. Obviously, if you believe that the trials completed cannot prove that the antibiotic was effective, any safety risk is logically unacceptable.

The FDA undertook its own analysis of the voluntary database that tracks physicians' reports of adverse events for marketed drugs. Their own analysis of the risk of liver toxicity caused by Ketek as determined by data mining of the reporting system database suggested that the compound was associated with no more risk than other, older antibiotics or than Tylenol. Serious liver toxicity from Ketek was estimated to occur 1 in 100,000 to 1 in 200,000 courses of therapy. To put this in perspective, fatal allergic reactions from the penicillins occur with a frequency of 1 in 50,000 to 1 in 67,000 courses of therapy, and serious reactions occur as often as 1 in 7,000 courses of therapy. Amoxicillin-clavulanic acid (Augmentin) (a penicillin analog) is the biggest selling antibiotic in history with peak year sales of around $2B. Augmentin is now generic. It also causes more cases of serious liver toxicity than any antibiotic on the market including Ketek. But the penicillin drugs including Augmentin are still frequently used to treat sinusitis and bronchitis with the continuing approval of the FDA. Tylenol is one of the most widely used drugs in the world (although not an antibiotic) and causes more cases of acute liver failure requiring liver transplant than any other drug. It is still sold without prescription worldwide. The advisory committee voted for withdrawal of marketing approval of Ketek for sinusitis and bronchitis at the meeting. The FDA followed their advice shortly after the 2006 meeting.

During the scandal leading up to the 2006 advisory committee meeting, Senator Grassley and Representative Markey were threatening a full-scale congressional investigation of the approval of Ketek. They seemed to suspect that the agency was somehow in cahoots with industry in general or with Aventis in particular or was simply incompetent. The FDA anti-infectives group went into defense mode and was working many extra hours and days supplying materials to congressional staffers. It is possible if not likely that the entire 2006 advisory committee meeting was a response to this political pressure. In fact, the pressure seemed to dissipate after the advisory committee meeting and with the reassignment of the Director of the Divison of Anti-Infective and Ophthalmology Products at the FDA.

Pneumonia – The New Frontier. New Trial Requirements for Pneumonia Will Make Approval Much More Difficult and Costly and Sometimes Simply Infeasible

Along the same lines as their inquiry on otitis, sinusitis and bronchitis, the FDA recently examined the role of antibiotics in pneumonia. Those of us in the infectious diseases community held our collective breath waiting to see if the FDA would decide that they did not understand whether antibiotics had an effect on bacterial pneumonia. To us clinicians, that antibiotics have a dramatic beneficial effect in the treatment of pneumonia was obvious and well proven by our own personal experiences as physicians and by clear historical precedent. Many of us could not understand what the FDA was thinking.

Pneumonia is generally divided into two categories – infections acquired in the community and those acquired while in the hospital. Both can be lethal. Community-acquired pneumonia strikes four to six million Americans every year. 600,000 are hospitalized and tens of thousands die leading to an annual cost to the US of over $10 billion.

In order to begin to understand the importance of antibiotics in the treatment of pneumonia, we have to delve back into history. When antibiotics were first developed in the 1930s (sulfa drugs) and the 1940s (penicillin), clinical trials as we know them today were not performed. Physicians would treat patients with the antibiotic and then search among hospital records for other, similar patients who were not treated (historical controls) or they would compare treated patients to similar patients in the hospital at the same time but who were not treated (concurrent controls). Obviously, this is not the same as asking a patient to agree to be either treated or not in a blinded fashion such that neither the patients nor the treating physicians know who is on which therapy as we would do in a contemporary trial. Nevertheless, when we look at deaths from pneumonia in these older studies, antibiotics prevented anywhere from 16 to 26% of them.

This analysis is not as straightforward as we would like. Community-acquired pneumonia has a variety of causes and not all of them are treated well with penicillin or sulfonamides, the antibiotics studied in the 1930s and 1940s. Some cases are actually due to viruses that are not treated by antibiotics at all. Even so, antibiotics still prevented 16% of deaths in all comers in those early studies. Table 4.1 below has been modified from the position paper presented by the Infectious Diseases Society of America to the FDA in 2008.

Table 4.1 Historical Studies of Antibiotics (penicillin or sulfonamides) in Patients with Pneumonia

	Untreated mortality	Treated mortality
Historical-control Studies	2184/5747 (38%)	398/3293 (12%)
Concurrent-control Studies	58/254 (23%)	21/308 (7%)

In other studies, if we look at only those patients who were infected with the most common bacterial cause of pneumonia, *Streptococcus pneumoniae*, and among those we look only at those patients that had invasion of the bloodstream by the bacteria (10–20% of patients), penicillin or sulfonamide prevented anywhere from 30 to 80% of deaths. If we don't look at deaths, since very few patients die from pneumonia today (because we have such good antibiotics in large part), but we rather look at time to normalization of temperature, we get a different perspective. In one study of sulfonamides from the 1930s, 84% of antibiotic treated patients were afebrile by 72 h compared to 2–3% of untreated patients. Clearly, antibiotics work and they work well. The overall mortality for pneumonia today, taking all comers in clinical trials, is around 3%. Any physician who has treated a patient with bacterial pneumonia can testify to the dramatic effect of antibiotics in this disease (see Chapter 2).

Do we need new antibiotics for community-acquired pneumonia? Probably not today. Our antibiotic armamentarium is diverse enough given the emergence of resistance that we are seeing that, in my own view, there is not an urgent need today for new antibiotics to treat this disease. The problem is that we have to think 7–15 years into the future. The risk is unpredictable. Who would have predicted the epidemic of vancomycin-resistant enterococci that has devastated our intensive care units since 1989? My crystal ball is telling me that in 10 years, having a new antibiotic for this disease would be worthwhile, especially for those patients allergic to the penicillin type drugs, given today's rate emergence of resistance to the non-penicillin type antibiotics. Of course, today if you are one of those rare individuals who can take neither the penicillins nor the quinolone antibiotics, you still have Ketek, which remains approved for pneumonia.

What has the FDA decided about community-acquired pneumonia? Who knows as of this writing? They have issued draft (I emphasize draft) guidelines that, in my view, will make it extremely difficult if not impossible to carry out clinical trials in this disease. First, they require that no prior antibiotic be given. Scientifically, this is a sound decision since even a single dose of antibiotic can have a beneficial effect in pneumonia. In one recent trial, an antibiotic that was later found to be ineffective for pneumonia because it is actually inactivated in the lung looked successful among patients that had up to 24 h of effective prior antibiotic therapy but was clearly much less active among patients who had no prior therapy. Therefore, any experimental drug given after an effective antibiotic, even after only one or two doses, might look better than it really is. In fact, in recent trials, about 40% of patients had received at least one dose of another antibiotic before being enrolled in the trial. If those 40% had to be replaced with those who had no prior antibiotic, the trial would take much longer. For more serious cases of pneumonia, such as those now required by the FDA, it will be even harder to find patients with no prior antibiotic at all. Further complicating this requirement is a quality measure used for hospital accreditation for continued participation in Medicare and other reimbursement plans. This quality measure requires that the first dose of antibiotic be given for pneumonia within 6 h of the patient first being seen in the health care facility. Clearly, this will make it even harder to find untreated patients to enroll in clinical trials of antibiotics for pneumonia. The best solution to this problem is to allow only one or two doses of a prior antibiotic before entering patients into a trial for a new treatment. That would preserve most of the scientific integrity of the study and still allow a reasonable rate of enrollment. The analysis at the end of the study can then be stratified looking separately at those who had received prior effective therapy and those who had not. This approach would be a balance between perfect science and reality – a concept that seems to have escaped the FDA.

Second, the number of patients required has skyrocketed into the thousands further increasing the cost and the time it will take to run such a trial. The FDA has required that the new antibiotic demonstrate that it works only in those patients that have a documented (by culture) bacterial infection. Overall, in America, the number of cases of pneumonia that are documented by isolating the infecting organism is an astounding 7.5% (Medicare data). In recent clinical trials where we make heroic efforts to identify the bacterial pathogen, the diagnosis rate varies from 20 to 35% with an average of about 25%. Given these numbers and the FDA's statistical requirements, to gain approval of an oral drug for pneumonia would require studying over 5000 patients. My estimate is that the trials alone would take more than 5 years and that the data might be obsolete by the time the FDA actually approved the drug. The Infectious Diseases Society and the FDA recently held a workshop to discuss the use of modern diagnostics to boost the proportion of cases where the pathogen could be identified. It is clear that this is not an attractive market for diagnostics companies and that the pathway to approval for such a diagnostic test is not straightforward. One diagnostic company executive openly questioned whether "the juice is worth the squeeze." The only other way forward offered by the FDA that was at all practical was to propose the use of an investigational diagnostic method that would have to be considered on a case be case basis and where the implications for how the company would be able to promote the drug after using such a diagnostic test was unclear. The company might be stuck having to promote their antibiotic to be used only after a diagnosis had been established with an unapproved investigational diagnostic test. The design proposed by the FDA is quite simply not feasible in today's world and no pharmaceutical company will undertake such a study.

Finally, the severity of illness to be treated has increased. There is some belief that patients with the mildest forms of pneumonia are more likely to get better without antibiotics, so the FDA would like to eliminate them as much as possible from clinical trials so that the antibiotic being investigated is appropriately challenged. The science behind this decision is, at best, controversial and this requirement will further slow trial enrollment rates and increase costs. Mild pneumonia tends to get worse without treatment.

Some of these FDA requirements may be scientifically justified, but they will help to assure that no or only very few new antibiotics will be developed for this important infection.

To make matters worse, Public Citizen has publicly called for a different endpoint altogether for studying community-acquired pneumonia – that of mortality. The proposal is scientifically based on the fact that the benefit of antibiotics in pneumonia was proven by looking at mortality rates in treated compared to untreated patients in the 1930s and 40s. This approach is completely irrelevant in the antibiotic age. The problem with this suggestion is that almost no one dies of pneumonia anymore because our therapy is so good – especially when considering the more common forms of pneumonia. Overall, for community-acquired pneumonia treated as an outpatient, the mortality ranges from 0.1 to 0.9%. For pneumonia that requires that the patient be hospitalized, it ranges from 0.9 to 26.7% – but those patients in the high mortality range are those admitted to the intensive care unit where antibiotics may not be much help anymore. These ICU patients usually have severe mechanical problems with their breathing from all the fluid in their lungs and other derangements. The overall mortality in trials of patients with community-acquired pneumonia is around 3%. So the proposal to use mortality as an endpoint for studying pneumonia would require enrolling from 7200 to over 100,000 patients in a trial. Again, this is a completely infeasible design.

PhRMA has an antimicrobial working group that has provided a response to the FDA's new proposed guidelines for clinical trial design in pneumonia. They have made suggestions that respond to the FDA's requirement for strong science and for microbiologically documented infection while providing for the feasibility of future trials. Because the industry is thought of as the devil incarnate in Washington and around the nation, politically, the FDA has no particular incentive to listen to them.

In fact, there was such a negative response to the FDA's draft guidelines for community-acquired pneumonia that they held yet another Anti-infectives Drug Advisory Committee (AIDAC) meeting in December of 2009. I was one of the presenters at the meeting. Prior to the meeting, I had submitted my presentation materials showing that the FDA had mandated clinical trial designs that were infeasible. I pointed out that this was, in a way, irresponsible and misleading. They seem to have heard that message because in their summary, they recognized that the issuance of guidance requiring infeasible trials was not acceptable. The FDA also recognized that mortality was not the only acceptable endpoint for a clinical trial of pneumonia and they agreed with the world of physicians who said that one could tell whether a patient was responding to antibiotics for pneumonia by day 3 of therapy. These concessions could be a major turning point in our discussions with the FDA. I (and others, too) have communicated a trial design strategy that would allow for feasible trials and still achieve everything the agency would like in pneumonia. But I have been disappointed so often in the past that I won't break out the champagne until I see new guidance with feasible trial design requirements.

The FDA Can Change Its Requirements After Completion of a Trial and then Require New Trials for Approval

Another really interesting development at the FDA is the policy that they can change their minds about previous agreements. In the old days, maybe even up to 7 years ago or so, if you had a discussion with the FDA on your trial design and they agreed with it in writing, they would not reject your data because they had come out with new policy requiring a different design. This is important. A company will pay about $30 million for a single Phase III antibiotic trial. To get approval, you need to run two such trials for each indication (skin infection, pneumonia, etc.). These trials usually take about 2 years to run and at least another 6–12 months for data analysis and submission of the dossier to regulatory agencies. Most companies plan on spending about $70 million on trials and other requirements to get approval for a single indication. Before putting that much money in play, companies like some reassurance that the trial design they are using will, if the study reaches its endpoints, lead to approval by the FDA. There is a process in place at the FDA (Special Protocoal Assessment or SPA) where sponsors can submit specific trial protocols and get written comments back from the FDA. This, in the past, was extremely valuable to all parties. The withdrawal of approval of Ketek for sinusitis and bronchitis because they did not run placebo controlled trials that they didn't know they needed to run is one example of how the exchange between industry and the FDA has become dysfunctional. In 2005, Advanced Life Sciences started Ph. III trials of their antibiotic, cethromycin, which they had licensed from Abbot a number of years earlier. It is similar to Ketek, but thought to have a better safety profile. They agreed a trial design with the agency and completed their trials in 2007. An NDA (New Drug Application for approval to market) was submitted to the FDA and accepted by them in 2008. In 2009, the FDA informed Advanced Life Sciences that the trial data submitted did not prove that their drug was efficacious because the design and therefore results did not conform to guidelines promulgated by the agency in 2009. Advanced Life Sciences is now in the throes of trying to complete such a trial. Their drug will now be delayed by years if it ever does get approved.

An even more abysmal example of this occurred just after Thanksgiving, 2009 when the FDA notified Theravance that the data from two phase III trials in hospital-acquired pneumonia would have to be analyzed based on a primary endpoint of all cause mortality and not the previously agreed primary endpoint of clinical outcome. Theravance had designed these two trials to look at clinical improvement in the actual pneumonia being treated as a primary endpoint and examined all cause mortality as a secondary endpoint as agreed with the FDA at the outset of the studies. If Theravance is unable to provide enough data to satisfy the FDA statistically that they succeeded in improving the new primary outcome of all cause mortality, they will have to run at least one more study in this indication. I would estimate that trials in hospital-acquired pneumonia probably run around $50 million each because of the complexity of the patients and the severity of the illness involved. So Theravance and its partner Astellas have probably already sunk $100 million into these trials and they may have to sink yet another $50 million to get to the new goalposts – and this will only be after at least another 2–3 years of study.

Another company, Replidyne, went belly-up following a similar change of opinion in midstream by the FDA. If companies cannot have some assurance that the large investment required up front for clinical trials of antibiotics won't be thrown down the toilet from the get-go because the FDA can change the goalposts at any time, why should they take the risk at all? This is especially true now when the goalposts seem to move every month!

The FDA is Regulating Itself Out of the Antibiotics Business

How did the FDA get where it is? The FDA started by trying to justify the statistical margins used to demonstrate that one antibiotic (the new one under study) was not inferior to the antibiotic used as a comparison – the so-called gold standard therapy. We have to design our trials that way because to use a placebo for patients with infections would, in most cases, be unethical at best and criminal at worst. The margin defines statistically how inferior in the worst case the new antibiotic might be. This does not mean that the new antibiotic might actually be that much inferior (10–15% in most modern studies) but that this is the limit beyond which the FDA will not go in approving the drug. So, in recent years, the FDA has been scratching their heads trying to find justifications for defining these margins. The idea is that the underlying assumption when you say that one antibiotic is not inferior to the other is that the standard to which you are comparing the new drug is still better than no antibiotic at all. Of course, to us infectious diseases physicians, this is patently obvious and we don't need statistics to tell us that antibiotics work. But apparently the FDA does. The FDA then searches for historical data coming from the days – 80 years ago – when antibiotics were studied in comparison to no therapy. The endpoint used in those studies was mortality. To be scientifically consistent, the FDA always heads back to these 80 year old studies and seems to want us to keep doing things the way they were done back then, even though 80 years have passed and both the data and the methods are no longer scientifically relevant. This must change if we are ever to make any progress in developing and approving new antibiotics.

Table 4.2 below shows indications and whether or not the trials required by the FDA are feasible or not. Also shown is my own view of the industry's opinion

Table 4.2 Indications for marketing approval available from the FDA vs. current feasibility of trials in those indications

Indication	Are trials feasible	Market attractive
Skin infections	*Yes*	*Yes*
Community-acquired pneumonia	*No?*	*Yes*
Hospital acquired pneumonia	*No? awaiting guidance*	*Yes*
Ventilator associated pneumonia	*Yes – awaiting guidance*	*Yes*
Intra-abdominal infections	*Yes*	*Moderate*
Urinary tract infections	Yes	No
Bone and joint infections	No	Maybe
Heart valve infections	No	No
Fever in neutropenic (patients with low blood counts post chemotherapy) patients[a]	*Yes*	*Yes*
Otitis media	No	Yes
Acute bacterial exacerbations of chronic bronchitis	No	Yes
Acute bacterial sinusitis	No	Yes
Pharyngitis (strep throat)	?	No

[a]May no longer be considered a valid indication in Europe

on whether the market opportunity for a given indication is reasonable. I was able to identify only five indications where trials are still feasible and where industry still considers the market opportunity to be reasonable and these are indicated in italics in the table. The FDA is now considering guidance on clinical trial design for serious skin infections and for pneumonia acquired in the hospital. For both of these types of infection, but most especially for hospital-acquired pneumonia, there is a desperate need for new antibiotics now. If the FDA makes decisions that limit our ability to bring antibiotics forward for these infections, we will all pay a high price. And, at that point, there will essentially be no infections for which we can actually develop new antibiotics. The industry will simply pull out altogether (if they have not already done so).

In addition to the loss of commercially interesting indications for antibiotics, there is the rapidly changing pharmaceutical marketplace to consider. We will examine this from the industry point of view again in the next chapter. But consider the following. The US used to account for over 50% of the worldwide pharmaceutical marketplace. In 2009, the IMS estimates that the US will account for a record low 40% for a variety of factors we will consider later. But, for the FDA, this means that they are now less relevant in the world market than they used to be.

So it is quite possible that the FDA regulators will successfully regulate themselves out of their jobs.

We Need Balance and Perspective from the FDA

Back in 2000, when I was working for Wyeth and we were first presenting our views on clinical trial design in antibiotics to the FDA, we asked for a balanced approach. We all agree that the clinical trials upon which approval are based have to be scientifically sound and that the antibiotics we study must be appropriately challenged, must show good activity in the infections we are studying and they must be safe. The balance is required in that the ultimate trial design that we agree upon must be achievable within the resources and budgets that we have for antibiotics. I personally would also like to see some balance in the application of new guidance between new drugs and older drugs. If there is a risk of toxicity that we are not willing to accept for a new drug, why should we accept it for a much more widely used (and therefore more dangerous) older drug? We obviously have a long way to go.

One of the things I learned early on in my career, the first time I participated in the conduct of a clinical trial, was that clinical trials have little to do with life as we know it. As a physician, when I am confronted with a patient who might have a serious infection, say pneumonia, I can't not treat him or her because they are so sick they might not live more than 4 or 5 days or their kidneys are not functioning well or because they might also be infected with HIV, the virus that causes AIDS. I treat the patient as best I can and hope. But in a clinical trial, such a patient would frequently not be studied. Why? There are lots of reasons. If the patient in fact died before the minimum length of therapy, the data could not be used since for either the experimental drug or the comparator. The treatment has to be long enough that we can evaluate whether it was successful or not. HIV infected patients might be harder to treat since they might not have a normal immune system. These all may be good reasons for excluding our hypothetical patient, but they also explain why trials are not real life.

Another disconnect between clinical trials and life on the wards is the fact that most, probably 70–80%, of initial antibiotic therapy is empiric. By that I mean that either the physician does not know the bacteria he/she is treating or, sometimes, doesn't even know the site of the infection (lung, urinary tract, etc). He/she just knows that the patient has a fever or other signs of infection that needs treatment. Usually, such a patient will be treated expectantly with an antibiotic or even more than one antibiotic to make sure that all the likely bacterial pathogens are treated. Only later, if a bacterial pathogen is identified or the site of infection declares itself, might the antibiotic therapy be more specifically tailored to the specific infection at hand. When I was consulting at the VA hospital, more often than not, if the patient had responded to the broad-spectrum therapy that was used initially, there was a great reluctance to change to a more specific therapy even when a specific pathogen had been identified. There is always a lingering doubt that maybe what was identified is not the entire answer. I frequently get the feeling that the FDA has lost sight of these issues and they actually believe that the trials they require reflect clinical practice. We will return to this subject in Chapter 7.

The FDA Makes It Difficult for Them to Obtain Good Advice

The FDA works closely with advisory committees. The FDA's Office of Antimicrobial Products has a Anti-Infective Drugs Advisory Committee (fondly known as AIDAC) that it uses to help its reviewing divisions with decisions around approvals, withdrawals, questions of trial design and other important issues that require public airing. There is also a strict conflict of interest policy that limits or bans the participation of individuals with a financial interest in the particular decision or decisions the committee confronts. The guidance actually speaks about a $50,000 limit that, in my view, is too generous. In its application of these policies I think the FDA has at times gone so far as to limit the expertise of its anti-infectives committee. The problem is that industry, like the FDA, needs outside advice. If folks in industry had no one to speak with other than themselves, I'm not sure we would ever have any products. These same experts can also provide valuable advice to the FDA. The anti-infectives world is a very small one and shrinking all the time. I once complained to the director of the anti-infectives reviewing division at FDA about the quality of the anti-infectives advisory committee. I felt that very few on the committee were experienced in antibiotics per se and especially in clinical trial conduct and design for antibiotics. These topics were almost always part of the committee discussions. She did not disagree with me and asked me for recommendations for new committee members. She said they could not work with people in industry outside the single non-voting member that is a part of every advisory committee. She wanted nominations for women, professionals of color, and people from regions other than the Northeast. As I considered the people I thought had the experience and expertise the FDA needed at the time, I came up with a list of names – almost all white men from the East coast. None were appointed.

This discussion reminds me of a recent set of news articles noting that Senator (retired) Tom Daschle works for a lobbying firm tied to the health insurance industry while he is simultaneously a close advisor to President Obama on health care reform. Is this an apparent conflict of interest? Sure. Is President Obama able to put this in perspective when discussing things with Mr. Daschle? Of course. FDA scientists and reviewers are not stupid, either. They know how to sift advice. As long as everyone discloses his or her potential conflicts, a conversation can take place.

The FDA still works with outside stakeholders such as PhRMA and the Infectious Diseases Society of America and others. Unfortunately, good advice from all is frequently ignored.

What has happened to the FDA since 2000? The FDA was leaderless for most of the last 9 years. Either the appointed commissioners did not last long or there was an acting commissioner for most of this time. In the anti-infectives group, there has been a significant loss of key individuals in leadership roles who understood the more practical problems of trial design and struggled in tandem with industry to achieve both the FDA's charge of assuring efficacy and safety of marketed products and industry's goals of doing so in a feasible way. With a new and dynamic commissioner, I hope we will see a more balanced approach to antibiotics such that we will be able to have them when we need them.

See the Chapter 7 for suggestions on possible ways forward for us all.

Chapter 5
Europe

The major pharmaceutical markets are the US, Europe (where the UK, Spain, France, Italy and Germany are the major players), Japan and the rest of the world. As I explain in the section on industry, the US comprises about 40–50% of the market, Europe with 20%, Japan with 10% and the rest of the world at 20%. As the US share drops because of the recession and increasing competition and health care reform, the emerging markets like China, India and others are taking up the slack. Europe, nevertheless remains an important market. They have a complex regulatory environment and tough pricing policies. Europe has been ahead of the curve in recognizing the perfect storm afflicting antibiotics. Whether they will actually act in accordance with their published analyses and resultant recommendations for dealing with the storm remains to be seen.

As part of the formation of Europe as a common market, the European Medicines Agency (EMEA) was established as the regulatory body for the group of nations. Prior to this, sponsors had to submit a separate dossier of data to each of the national regulatory agencies within the European block to obtain marketing approval for each individual country. This approach is still possible. But with the establishment of the EMEA, sponsors now have the option, and for some drugs the requirement of submitting a single dossier to Europe (Centralized Procedure) and gaining marketing approval for the entire block of nations. The actual opinions of the EMEA are provided by the Committee for Medicinal Products for Human Use or CHMP. Within the CHMP are Scientific Advisory Groups which provide advice to both sponsors and the CHMP regarding products in key areas. There is a specific advisory group for anti-infectives (antibiotics and anti-fungal compounds). The approval by the EMEA with all its CHMP caveats must be followed by a negotiation with each individual country as to the terms under which the product can be marketed within that nation. Each country maintains a national negotiation for establishing conditions of marketing and product price. This is a completely different situation than that found in the US where there is no national negotiation at all. There are large blocks of patients that negotiate price, such as the Veterans Affairs administration and others, but there is no truly national price negotiation. This accounts for the fact that drug prices in the US are consistently higher than they are in other countries.

D.M. Shlaes, *Antibiotics*, DOI 10.1007/978-90-481-9057-7_5,

Europe also has much tighter controls on how antibiotics are used. Each country sets criteria under which the use of an antibiotic is reimbursed either to the hospital or to the patient. Any use outside of these pre-set criteria means that the government is no longer obligated to pay for the prescription. This system gives the European nations much more control to prevent abuse of antibiotics and, accordingly, it limits the market for antibiotics. Thus, while Europe has 30% more population than the US, its pharmaceutical market is smaller – partly because of price and partly related to marketing restrictions within each of the member nations.

One of the great advantages of the FDA is that it is easy to speak with them throughout development starting at the earliest stages and going all the way through to post-marketing activities. Europe is not constructed that way. To speak to the EMEA, you must apply for scientific advice. In this case, the Scientific Advisory Group is convened and the sponsor's questions are dealt with in a formal meeting. This meeting takes around 12 weeks to set up. The responses by both sponsor and EMEA are said to be binding. But, like many binding contracts, if things change, which they often do, the agreements are less solid. Nevertheless, the EMEA has been in general more bound by these agreements than the FDA. The FDA has made it clear in recent years that it no longer feels bound by such agreements and they frequently change their goalposts in mid-stream. Europe seems still much less likely to do that. Nevertheless, speaking to Europe is much more cumbersome. One option for sponsors is to speak to different national authorities like the UK, France or Germany. I personally prefer this to the formal advice route with the EMEA. The disadvantage to this is that you are not speaking to the group to whom you will ultimately submit your dossier.

Europe has been much more forward thinking than the US regarding the dilemma now facing antibiotics and the pharmaceutical industry. With the caveat that thinking is one thing and acting is another, we should explore what Europe has been doing, as it could be a model for all of us. Sweden, which assumed the rotating European Presidency during 2009, has been particularly active in this regard and just held a large meeting on this subject in September where the Obama administration was represented. Three documents have been published and are available on the web for those that are truly interested; *Policies and incentives for promoting innovation in antibiotic research*, commissioned by the Swedish Government and written by Professor Elias Mossialos and his co-workers at the London School of Economics and Political Science on behalf of the European Observatory for Health Systems and Policies; and *The bacterial challenge: Time to react. A call to narrow the gap between multidrug-resistant bacteria in the EU and the development of new antibacterial agents,* jointly written by two European agencies, the European Centre for Disease Prevention and Control (ECDC) and the European Medicines Agency (EMEA) in collaboration with the international network Action on Antibiotic Resistance, ReAct. The resulting report from the meeting held in Sweden in September, 2009 is also available. It is entitled, Innovation Incentives for Effective Antibacterials.

In the analysis by Europe, they estimated that 25,000 European patients died from an infection caused by one of the ESKAPE organisms noted in the Chapter 3.

They estimated that these infections led to 2.5 million additional hospital days and additional in hospital costs of 900 million Euro ($13.5 billion) per year. Overall, the additional costs to society of these infections was estimated at 1.5 billion Euro ($2.25 billion) per year. In their report, it was clearly stated that these figures almost certainly represent an underestimate since all costs, such as those attributed to intensive care for example, could not be taken into account in their model.

At the same time, as shown below (Fig. 5.1), they noted that there were only 15 antibiotics under development by pharmaceutical companies and biotech that might be useful in treating infections caused by the ESKAPE organisms. Of these, only eight were active against the Gram negative members of ESKAPE where there is the greatest perceived medical need. Of the total 15 compounds, three have already completed trials and have been filed – all active against Gram positives. Of the eight compounds active against Gram negative pathogens only four have survived to Phase II or later trials and none has yet been filed. Given the overall chances of being approved in our current times, it is likely that few of these products in development will make it all the way to market. The European analysts clearly recognize this risk.

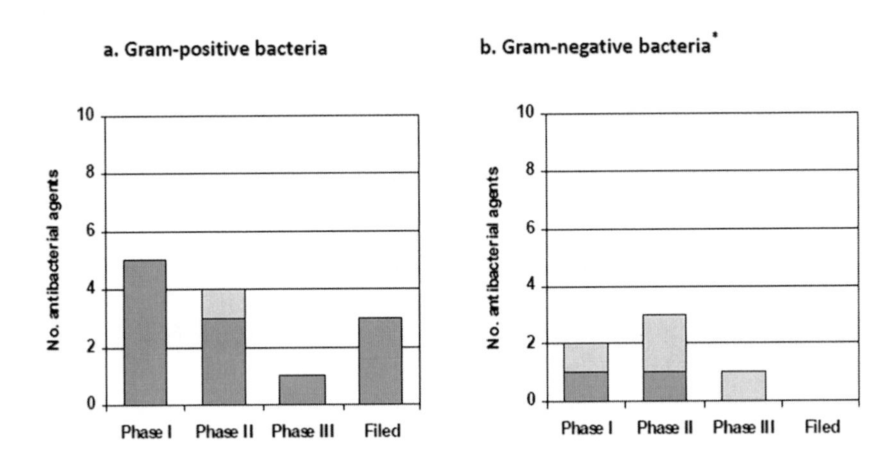

Note: In vitro activity based on actual data is depicted at the bottom of each column in darker colour. Assumed in vitro activity based on class properties or mechanisms of action (where applicable) is depicted in a lighter colour at the top of each column.

** Two carbapenems have been omitted from Figure 4.4.5b since they are no more active than earlier carbapenems against Gram-negative bacteria. The relative novelty of these agents was based on a better profile of activity against antibiotic-resistant Gram-positive bacteria and are therefore included in Figure 4.4.5a.*

Fig. 5.1 New systemic antibacterial agents with a new target or new mechanism of action and in vitro activity based on actual data (*dark colour bars*) or assumed in vitro activity based on class properties or mechanisms of action (*light colour bars*) against the selected bacteria (best-case scenario), by phase of development ($n = 15$). From Mossialos et al. (2008), a publicly available report from the London School of Economics and the European Observatory on Health Systems and Policies

They also examined regulatory issues in antibiotic development in Europe. Key stumbling blocks were around clinical trial design – as they are in the US with the FDA – and drug pricing policies in individual member states. Although the regulatory authorities in Europe have said that they will be more flexible than their FDA colleagues, this has yet to be seen by the industry. Further, with the continued US dominance of the marketplace, if the FDA insists on infeasible trial designs, it might not matter what Europe does. Changing European regulatory guidelines to be more accommodating to antibiotic development in the absence of similar changes in the US will only work if the US market dominance recedes or if the FDA can be dragged along with the EMEA. It is not clear if the EMEA is actually in synch with the desires expressed in this regard by their EU Commission colleagues. For example, in the alterations to European guidelines for the development of antibiotics, it is clearly stated that they will also prefer placebo-controlled trials for self-limited infections such as otitis, sinusitis and bronchitis. Some European authorities with whom I have communicated state that they will be more flexible here than is suggested by the guidelines. But we will only know this if someone actually proposes a new trial in one of these indications. At this time, that seems highly unlikely given the written perspectives of Europe and the stated guideline in the US.

Europe also carried out a very detailed investigation of the economics of drug discovery and development in order to understand which incentives might work best for industry. These will be considered later in the Chapter 7. Nevertheless, the entire analysis was presented in international meetings on the antibiotic dilemma convened by Sweden in late 2009. The result of these studies and meetings was a recommendation to the Council of the European Union to develop and implement strategies as noted. The resulting resolution by the Council is cited in its entirety below. This result shows that Europe is very clearly ahead of the US and the rest of the world in its thinking on antibiotic discovery and development.

The Council of the European Union met on November 30-December 1, 2009 and adopted the following resolution for the European Union:

CALLS UPON THE MEMBER STATES TO

– develop and implement strategies to ensure awareness among the public and health professionals of the threat of antibiotic resistance and of the measures available to counter the problem;

– ensure the development and use of integrated strategies to diminish the development and spread of antibiotic resistance as well as healthcare-associated infections and their consequences, encourage healthcare institutions to have structures in place as well as ensuring effective coordination of programmes focusing on diagnosis, antibiotic stewardship and infection control;

– review and consider options to strengthen incentives to conduct research and development of new effective antibiotics within the academic as well as the pharmaceutical sector as a whole, taking into account the situation of small and medium-sized enterprises. These options and methods could include cost-effective

push mechanisms to remove bottlenecks in the early stages of research and development of new antibiotics and pull mechanisms to promote the successful introduction of new products.

15. CALLS UPON THE MEMBER STATES AND THE COMMISSION TO

– support the sharing of research infrastructures, recruitment of researchers, stimulation of and support for global research cooperation, increasing the spread of research results and knowledge through information exchange structures and considering existing and new financial instruments;

– explore ways to promote further public-private partnerships between industry, academia, non-profit organisations and the healthcare system to facilitate research into new antibiotics, strategies for use of currently available antibiotics and diagnostic methods;

– within the legal framework for market authorization of medicines, facilitate development of new antibiotics for which a particular medical need exists and when only limited clinical data can be submitted by the applicant for objective reasons, take full advantage of additional means of assessing safety and efficacy such as the utilization of preclinical assessment tools and pharmacokinetic data analyses;

– identify appropriate regulatory instruments to facilitate early approval for new antibiotics for which a particular medical need exists, in terms of providing continuous EMEA and national competent-authorities-assisted scientific advice, including strategies for adequate post-authorisation follow-up with an emphasis on safety aspects, including monitoring of antibiotic resistance;

– examine how to keep effective antibiotics on the market;

– while facilitating the development of new effective antibiotics, ensure prevention of healthcare-associated and other infections as well as the rational use of existing and new medicines;

– ensure that all actions are appropriately co-ordinated between different stakeholders from the sectors involved, such as health, finance, economic, legal and research.

16. CALLS UPON THE COMMISSION TO

– within 24 months, develop a comprehensive action-plan, with concrete proposals concerning incentives to develop new effective antibiotics, including ways to secure their rational use; and ensure that these proposals take account of the economic impact on the financial sustainability of healthcare systems.

– consider using experience regarding relevant procedures from previous specific EU legislation on orphan drugs and drugs for paediatric use to stimulate the development of new antibiotics for which a particular medical need exists;

– monitor and regularly report back to the Council on the public health need for new antibiotics, based on the emergence of antibiotic resistance, the characterisation of

new resistant pathogens and new antibiotic medicinal products and other methods to treat and prevent infectious disease in development as well as to propose further action when appropriate.

It is clear the Europe has expended an extraordinary amount of energy in their consideration of the problem of antibiotic resistance and the dry antibiotic pipeline. They have examined a range of solutions many of which are sensible. The problem is, that I do not see that they can be implemented in any kind of reasonable time horizon.

The economic solutions suggested consist of so called hybrid push pull mechanisms. For a detailed discussion of this issue, see Chapter 7. But, in brief, the example they give is one where a governmental body – Europe in this instance – purchases an "option right" to purchase a supply of a new antibiotic at some early stage in its development. This would be similar to the partnering activity that goes on now between small and large pharmaceutical companies where the large company purchases rights to market the small company's products in return from some monetary consideration and usually an obligation to pay for future development. In the case of the European Commission, the "option right" would provide funding for further development thereby reducing the risk for the company. This is the so-called push mechanism. The additional carrot, or "pull" mechanism, would be the obligation to purchase a certain amount of the product if the company is successful in winning marketing approval through the usual regulatory process. The problem I see with this is simply one of cost. The option right would cost on the order of $100 million to offset a company's development costs and risk. The guaranteed purchase would have to be considerable – on the order of $500 million or more. In return for this, the government realizes no monetary gain directly. On the other hand, they do save considerable money on health care overall if the antibiotic is active against resistant strains that are currently costing the governments on the order of hundreds of millions of dollars per year. While this kind of hybrid mechanism might work, the so-called wild-card patent extension discussed in more detail in Chapter 7 would probably work better.

Chapter 6
The Industry

Antibiotics Form a Cornerstone of the Pharmaceutical Industry

The pharmaceutical industry has its origins in the folk remedies of ancient times. Our first historical records of this activity come from the ancient civilizations of Asia. Apothecaries were established during medieval times especially in Europe. Most of the modern pharmaceutical companies or their precursors were established in the nineteenth century. These small companies and even individual producers commercialized, to various levels of scale, products thought to benefit health in one way or another. Some of the nostrums and remedies peddled during the nineteenth century were outright poisons and it was these that ultimately led to the formation of the FDA. The regulatory requirements plus those of a demanding public led the industry to establish rigorous manufacturing practices. Later, the advance of science led to new vaccines and chemical products that were valuable for physicians, patients and for the emerging industry. These events led directly to industry's turn towards scientific research as the basis for discovering new products in the early twentieth century.

The great thing about the pharmaceutical industry is that it is driven by science and it is tied to medical need. Two of the key early products from science driven discovery in industry were sulfonamides (1932) and penicillin (1943) discussed in Chapter 2. Thus, antibiotics became a cornerstone of the modern pharmaceutical industry.

Figure 6.1 and Table 6.1 below depict the timeline of the discovery of "new classes" of antibiotics and their later approval by the FDA. The figure and table serve to identify two issues. First, entirely new types of antibiotics are hard to find. Of those that are found, which have been many, only a tiny fraction make it all the way to the marketplace. Only two new classes of antibiotics (oxazolidinones e.g. linezolid or Zyvox and lipopeptides e.g. daptomycin or Cubicin) have been discovered and marketed in the 40 years following trimethoprim (a component of Bactrim) in 1968. Second, many of the antibiotics we have today would be called "me too" drugs, another very misleading characterization. I would like to spend some time placing this second issue in some context.

D.M. Shlaes, *Antibiotics*, DOI 10.1007/978-90-481-9057-7_6,

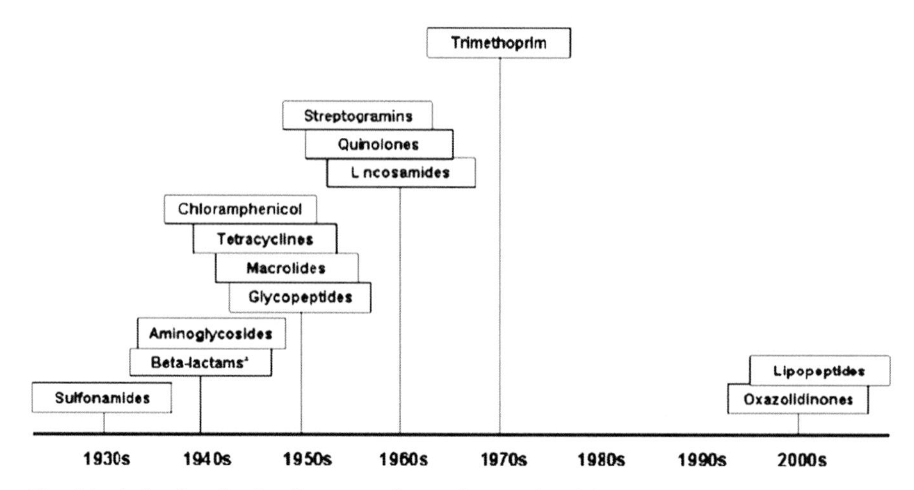

Fig. 6.1 A timeline for the discovery of new classes of antibiotics. Taken from The Bacterial Challenge (2009)

Table 6.1 A history of antibiotic approvals in the US

History of antibiotic approval in the US	
New antibiotic class	Year approved
Sulfonamides	1935
Penicillin	1941
Aminoglycosides	1944
Cephalosporins	1945
Chloramphenicol	1949
Tetracyclines	1950
Macrolides/lincosamides/streptgramins	1952
Glycopeptides	1956
Rifamycins	1957
Nitroimidazoles	1959
Quinolones	1962
Trimethoprim	1968
Oxazolidinones	2000
Lipopeptides	2003

The so-called Golden Age of antibiotic discovery and development probably occurred between 1955 and 1985. During these years, over 100 compounds entered into clinical trials and sixty eventually made it to market. The vast majority of these were derivatives of older classes, especially the cephalosporins, which are similar to the penicillins. The cephalosporins were more readily accessible to chemical modification, which in turn led to the discovery of new versions. Was there a medical need that drove the companies to explore the cephalosporins? Absolutely.

The early compounds like cephalosporin C and N had poor antibacterial activity compared to penicillin but were active against some of the penicillin-resistant strains of the time. The emergence of penicillin-resistant strains of staph was getting serious in the 1950s especially in hospitals in the US. Cephalosporins also appeared to have better activity against the Gram negative organisms (broader spectrum of activity) than penicillin did. Therefore, there was the promise that clinicians could have an antibiotic, possibly with the safety of and exquisite activity of penicillin, but with a spectrum of activity against more different types of bacteria. Tetracycline is also active against a broad spectrum of bacteria but it had problems. It could cause kidney and liver toxicity and it discolored the teeth and bones of children under age 8. The first broad-spectrum cephalosporins to enter the market were cephalothin and cephaloridine. The latter turned out to be somewhat toxic to the kidneys and never achieved the success of cephalothin. Cephalothin was an extremely important addition to the antibiotic armamentarium because it was active against penicillin-resistant strains of staph and a large variety of Gram-negative pathogens like *E. coli, Salmonella,* gonococci, and others. Both of these early cephalosporins were only available for intravenous administration and had to be administered several times per day. For the next 40 years, scientists would endeavor to improve the spectrum of activity of the cephalopsorins, to decrease the number of doses required and to make them useful when administered orally. These efforts were largely successful leading to several generations of cephalosporins with ever increasing activities against more bacteria and with simpler dosing regimes. We now have four generations of cephalosporins to choose from. Of course, bacterial resistance has been up to this challenge as well.

Another set of so-called "me too" antibiotics were discovered and moved through trials to approval. They were those containing Beta-lactamase inhibitors. These are second molecules, added in fixed combination to certain penicillins. Bacteria become resistant to penicillins and cephalosporins with an enzyme that destroys the warhead (the Beta-lactam ring) of the antibiotics – appropriately called Beta-lactamase. The inhibitors discovered by industry specifically target these enzymes and prevent the destruction of the antibiotic allowing it to find its target in the bacteria and kill them. Two of the best-selling antibiotics of our time are examples of this innovation – amoxicillin + clavulanic acid (Augmentin) and piperacillin + tazobactam (Zosyn).

The first aminoglycoside to have been discovered and marketed was streptomycin. It has mainly been used in the therapy of TB, and early on, in certain very serious bacterial infections of the heart. An entire series of new aminoglycosides with improved activity against resistant bacteria were discovered and marketed during this heyday of antibiotics. With their improvements in antibacterial spectrum and sometimes even in toxicity, the aminoglycosides could not be called "me too" drugs when compared with streptomycin. On the other hand, outside of urinary tract infections, the aminoglycosides did not work will when used alone. So they were often combined with other antibiotics for the treatment of more serious infections. For a while, they fell out of favor somewhat because of their toxicity to kidneys and the hearing apparatus. As bacteria become more resistant and difficult to treat, we have seen a bit of resurgence in their use.

When staphylococci became resistant to penicillin, industry answered with another "me too" penicillin called methicillin in addition to the cephalsosporins noted above. The so-called methicillin-resistant *Staphylococcus aureus* or MRSA, as these strains were designated, first emerged in the 1960s shortly after introduction of methicillin. They were also resistant to the cephalosporins. The glycopeptides were exemplified by vancomycin, introduced just in time for the discovery of MRSA against which vancomycin had good activity.

In addition to the explosion in new and better cepahlosporins, beta-lactamase inhibitors and the aminoglycosides, the industry also discovered and brought to market five entirely new classes of antibiotics including vancomycin as noted in the table. The rifamycins have found more use in the field of TB than as an antibacterial per se. The discovery of the quinolones with nalidixic acid in the early 1960s ultimately led to the discovery of the fluoroquinolones like ciprofloxacin and levofloxacin that have become mainstays of our antibacterial armamentarium today. Finally, trimethoprim became an important part of a fixed combination with an old sulfonamide, sulfamethoxazole to become Bactrim and Septra, which are still widely used today. This combination antibiotic combines to inhibitors of different steps in the biochemical pathway bacteria use to make folic acid, which they must have to survive. The dual antibiotic combination is particularly lethal. Unfortunately, as for most of these older antibiotics, resistance has all but finished the usefulness of this drug combination.

Resistance Should Create Opportunities

You would think that the emergence of antibiotic resistance would make it obvious to physicians that new antibiotics were needed, and that this need would result in brisk sales for any antibiotic that worked against resistant strains. Well its not so simple. The graph below (Fig. 6.2) illustrates the tremendous boost that the epidemic of MRSA provided the market for vancomycin. It also shows the beginning of vancomycin-resistance.

The data stops after 1996 when vancomycin became a generic and Lilly stopped promoting its branded product. But it is clear that the emergence of resistant strain where vancomycin was frequently the only choice left for seriously ill patients led to a tremendous market. A similar story can be told for the carbapenems where they have become the only reliable antibiotic for multiply-resistant Gram negative infections. The market for them has also increased dramatically over the last decade. Even these examples, where the market size varies from 400 to $800 million in peak year sales has not been enough to keep the industry engaged in the field of antibiotics.

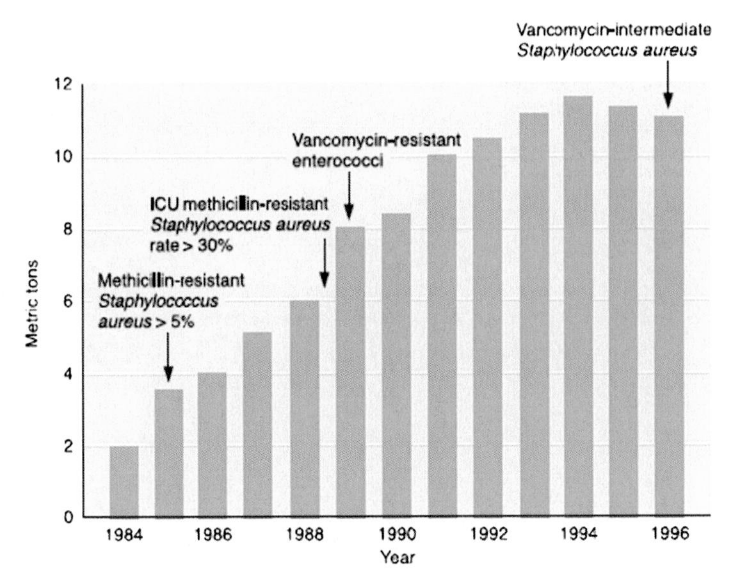

Fig. 6.2 Resistance creates markets and use creates resistance. Data on vancomycin use in the US 1984–1996. From Shlaes et al. (2004)

Industry Consolidation Reduces the Effort in Antibiotics

During the Golden Age, most pharmaceutical companies, large and small, were actively pursuing new antibiotics. Table 6.2 below shows the number of companies pursuing antibiotic research in 1990 compared to today. Back in the heyday, many in the industry did not consider that you were a pharmaceutical company unless you were active in antibiotic research and development. Based on my memory and that of Steve Projan (now heading Infectious Diseases research at Novartis), 18 mid to large size pharmaceutical companies were actively pursuing antibiotic research in 1990. This was the vast majority of such companies in existence. It was hard for us to identify any of the companies existing in 1990 that were not active in this area. Today the story is completely different. Of the 12 large pharmaceutical companies we identified, only four or five seem to be active in antibiotics research. I'm no longer sure about Merck since they recently reorganized.

What happened? First of all, over the last 30 years there has been tremendous consolidation within the pharmaceutical industry. As shown in Table 6.3 below, in 2003 Karen Bush of Johnson and Johnson looked at the histories of six of the large pharmaceutical companies over the preceding 20 years. Karen found that these six companies derived from mergers and acquisitions of 70 precedent companies. If extrapolated to large pharmaceutical companies in general, this would indicate a

Table 6.2 Evolution of antibacterial research in large pharmaceutical companies since 1990

Large pharmaceutical companies active in antibacterial research in 1990	Companies active today	Unknown	Companies not pursuing antibacterial research today
Abbott	Pfizer-Wyeth	Merck-SP	Abbott
Bayer	Astra-Zeneca		Bayer
Bristol Meyers	Glaxo SmithKline		Bristol Meyers Squibb
Ciba	Novartis		Lilly
Glaxo			Roche
Hoechst			J&J
J&J			Sanofi-Aventis
Lederle			
Marion Merrell Dow			
Merck			
Parke-Davis			
Pfizer			
Roche			
Rhone Poulenc			
SmithKline Beecham			
Squibb			
Upjohn			
Zeneca			

Table 6.3 Consolidation within the pharmaceutical industry 1980–2003

2003 Pharmaceutical company	Number of original companies since 1980
Aventis[a]	17
Bristol-Meyers-Squibb	8
Glaxo Smith Kline	12
Novartis	7
Pfizer	12
Wyeth	14

[a]Now Sanofi-Aventis
Pfizer has now purchased Wyeth
Not shown is Merck that has now purchased Schering Plough

91% consolidation over 20 years. Since then, we should add the merger of Aventis with Sanofi to form Sanofi-Aventis, and the recent purchases of Schering-Plough by Merck and of Wyeth by Pfizer. This means that there are simply fewer and fewer companies around who might be doing antibiotic research even if they were so motivated.

Antibiotics Are Not Financially Attractive in the Consolidated Industry

The companies remaining after this massive consolidation were much larger. This means that to drive an increase in their bottom lines from year to year, their revenues had to increase by much larger amounts. For example, lets take the giant that is about to be formed with the purchase of Wyeth by Pfizer. Pfizer had $48.3 billion in revenues in 2008. Wyeth earned $22.8 billion. The combined company will therefore have revenues of something on the order of $60B. These revenues will have to grow for the company to be successful. One probably can't cost-cut one's way to growth forever. If they want to grow the company by an extremely modest 5% per year, they will have to generate something like $3B in additional revenues. $3B does not grow on trees. It also does not usually come from antibiotics. Peak year sales for the largest selling antibiotic in history (Augmentin) were just under $2B dollars. I can count $1B antibiotics on one hand. Most anti-depressive drugs and most if not all the statins have peak year sales of over $1B before they lose exclusivity and generics invade their markets. Lipitor alone had peak year sales of $6.2B. Another way to make this comparison is shown in the able below. The data come from the early 1990s, but the principle has not changed. The data, from the Tufts Center for Drug Development Research, shows the value of various drugs over the life of the drug. While antibiotics are kind of average in value, they are dwarfed by the statins and the anti-depressants (Table 6.4).

Table 6.4 Net present value (lifetime earnings minus lifetime costs) of drugs

Net Present Value (NPV) of Drugs 1990–1994	
	Mean NPV
All Drugs	$0.8B
Antibiotics	$1.1B
Statins	$15B
SSRI anti-depressants	$11B

Another problem for antibiotics is, that given the time it takes to get from the lab bench to the market, companies are forced to try and look into their crystal balls anywhere from seven to 15 years in the future. Anyone who has worked in the field of infectious diseases can tell you how perilous that can be. While many predicted that we would have rampant epidemics of vancomycin-resistant staph in our hospitals, which never occurred, no one predicted the current rapid emergence of MRSA in our communities.

When these large companies examine their research portfolios and they compare the potential gain for an antibiotic with that for an anti-depressive, a statin, a drug for Alzheimer's disease or drugs in many other therapeutic areas, they have, more and more, chosen to deprioritize the antibiotics. That is exactly what happened at Wyeth when the FDA wanted to increase the stringency of the clinical trials required for approval of Wyeth's investigational antibiotic, tigecycline. Wyeth looked at the extra cost involved, compared the potential sales from tigecycline with other products in its pipeline and they just said no. Many companies, ultimately including Wyeth, have simply abandoned antibiotic research altogether as a way of prioritizing their efforts towards more profitable ventures.

One disadvantage for antibiotics is that in order for them to reach their full commercial potential, they must be developed in multiple indications each requiring large clinical trials. For instance, an antibiotic would undergo two trials in skin infections, two trials in community-acquired pneumonia, one trial in hospital-acquired pneumonia, two trials in fever associated with cancer chemotherapy, etc. In spite of this requirement, it used to be less expensive to develop antibiotics because the patients were only treated for 1–2 weeks instead of a year or more for chronic diseases. Drugs for high blood pressure might only be studied in that single indication, but they would be studied for a year or more in a large number of patients. That meant that even if the antibiotic would not pull in as much money, the up front expense of the trials would be less and the ultimate profits would be almost as good. But with the FDA's ever increasing trial requirements, those days are long gone. According to data from the Tufts Center for the Study of Drug Development, anti-infectives, which include antivirals, anti-fungals and antibiotics, are now almost twice as expensive to develop as drugs in other categories. The Center notes that these high costs are driven largely by antivirals, but I can add that the trend in antibiotics is the same.

Another advantage that antibiotics used to have is that the risk associated with them was lower. That is, overall, once a drug enters the first phases of clinical trials, it has only a 20% of ultimately being approved. Thirty percent of antibiotics used to make it. This lower risk is related to the great predictability of antibiotics based on test tube and animal model work described in Chapter 4. While it is still true that antibiotics have a somewhat better chance than other drugs at the beginning, their chances of failure in the very latest and most expensive stages of drug development have increased along with increasing FDA requirements such that antibiotics now have little overall improvement in their risk profiles compared to other drugs.

Therefore, much of the justification for antibiotic research that we used to tout to our company management, including the lower cost of trials and the lower risk associated with antibiotic development is simply no longer true. If that weren't enough, actually discovering new antibiotics is also a lot harder than it used to be. In the heyday, those decades from the 50s through the 80s, we could frequently find new antibiotics by screening microorganisms from soil samples and plants for anti-bacterial activity. But as the years went on, finding anything new was harder and harder. Why? Because the pharmaceutical company engines and academic labs around the world were good and had already found most of the antibiotics that you could find from this very simple method of screening.

Industry Takes a Risk-Averse Approach to Its Clinical Trials, but Deprives Us of the Most Important Information

While I think that the pressure on those companies still committed to antibiotic research is enormous, I feel compelled to pile on! The example I am going to use for this critique comes from clinical trials conducted by industry for drugs to treat resistant Gram positive infections like MRSA. As you have read earlier in the book, vancomycin has been a real workhorse antibiotic for the treatment of MRSA infections in our hospitals. At the same time, there have been a number of scientific and clinical studies to suggest that vancomycin isn't such a great antibiotic for the treatment of MRSA infections. Both linezolid and daptomycin were compared, in their late stage clinical trials, to vancomycin and both were shown to be "non-inferior." Thus, they both were no worse than vancomycin within a reasonable statistical margin of error. Because antibiotics work so well in a normal clinical trial, it is usually impossible to actually show superiority of one antibiotic over another. In a few, more specialized trials, there is some evidence that linezolid might be superior to vancomycin, but these data were never approved by the FDA. Had the FDA actually accepted these data and approved a label suggesting that linezolid was superior to vancomycin, Pfizer would have been allowed to make that claim in their marketing efforts. At the time that daptomycin and linezolid were studied, there was no antibiotic other than vancomycin for them to use as a comparator. But for telavancin, which was just approved in 2009, both linezolid and daptomycin had been approved and on the market for several years before the pivotal telavancin trials were started. Cerexa/Forest is developing a cephalosporin antibiotic active against MRSA. For their recently completed Ph. III trials, they also chose to use vancomycin as their comparator.

So why did Theravance-Astellas and Cerexa/Forest choose vancomycin as the comparator? I can think of a number of reasons why they should not have done so. First, vancomycin sells at a lower price. In Europe, the price of your comparator comes into consideration when you get into the ultimate price negotiations that you have to go through with each individual country to market a new drug. So using a higher priced comparator would have made the argument that you too can charge a higher price more palatable to the national authorities in Europe. In addition, some influential clinicians believe strongly that vancomycin is an inferior drug compared to both linezolid and daptomycin. There is some evidence to support this view even though the data could not be used by regulatory authorities to approve such a claim. In that case, why would you not want to show non-inferiority compared to what might be an inferior antibiotic? It is of interest that Cerexa/Forest are running a Phase II trial comparing ceftaroline to linezolid in the treatment of skin infections. This suggests that they are aware that some clinicians consider linezolid a superior drug but made an active decision not to pursue linezolid as a comparator in their first pivotal trials.

I think the answer to this question goes back to risk. Companies are risk averse and they worry that they might be wrong and that they might not succeed against the newer antibiotics even though the risk, in my view, is a small one given everything we know about all these drugs. I'm sure other considerations come into the discussion as well. Both daptomycin and linezolid are expensive. The companies, in this example Theravance-Astellas and Cerexa/Forest, would have to pay the going rate for the comparator drug for hundreds of patients. This would significantly increase the cost of the trials. Also, they might have considered that because vancomycin had been used in so many trials as a comparator, it was much better characterized and provided a more reliable comparator. Finally, they might not have wanted to provide additional data for their future competitors, Cubist (daptomycin) and Pfizer (linezolid). Perhaps they were all dissuaded by Arpida who carried out two Ph. III trials of their antibiotic, iclaprim, compared to linezolid and failed to show non-inferiority.

Theravance-Astellas or Cerexa/Forest might argue that, based on volume of use, vancomycin is still the most widely used antibiotic for the treatment of serious Gram positive infections in hospitals where MRSA is common. That vancomycin is the most widely used of the anti-MRSA drugs remains true partly for cost reasons, where vancomycin is considerably less expensive than linezolid and daptomycin. The fact that there are no data approved by a regulatory authority recognizing the superiority of either daptomycin or linezolid over vancomycin in the treatment of MRSA infections probably also contributes to this situation.

I believe that companies should be providing the best data for physicians and patients. This should be a prime consideration in the choice of a comparator for pivotal clinical trials. The comparator should be the gold standard. The gold standard is not always the same as the so-called standard of care.

Discovery of New Antibiotics Is Becoming Harder

In the 90s came the genome evolution. Not only could we decipher the human genome, but given that they were much smaller, we could more easily and more quickly decode the genomes of bacteria. Why was this important? Well, we could discover what makes them tick and how to kill them more efficiently. We could also, by comparing their genome with that of more advanced organisms, know which genes were unique to bacteria. Not only could we understand how to kill bacteria, but we could do so while not killing everything else, including people. We could apply advanced technology to all this. We could take the giant chemical libraries existing in large pharmaceutical companies and screen hundreds of thousands to millions of compounds against these bacterial targets. We could actually visualize the structure of the bacterial proteins involved using X-ray crystallography and other methods. We could then identify how inhibitors of these proteins might bind, and, using chemistry, improve the activities of the inhibitors and make them even more specific for bacteria. Or so we thought. The entire industry entered into a race to see who could sequence the most bacterial genomes the fastest, identify targets for killing the bacteria and then find drugs that hit those targets. What happened? So far, everyone, and I mean everyone, lost.

Glaxo-Smith Kline published the most widely known example of this scientific failure, but their results were typical of the entire industry. At GSK between 1995 and 2001, 67 screening campaigns on antibacterial targets were run against the SmithKline Beecham 260–530,000 compound collection. Only five chemical molecules were thought worthy of pursuing and none of those resulted in a drug. My own opinion is that once we figure out how to use the genomic information appropriately, we will be able to use it to find new antibiotics. That day is coming, but I don't think that we're there yet.

In 2008, the Pharmaceutical Research and Manufacturers of America (PhRMA) reported that their industry was sponsoring almost 22,000 clinical trials with drugs for cancer being the most common (almost 7,000 trials). PhRMA listed no trials for antibiotics (even though there are a few going on). This should give you an idea as to the priority industry is giving to antibiotics these days.

In summary, antibiotics are expensive to develop and getting more so. They are not less risky than other drugs overall and they are getting harder and harder to find. In addition, they are at best only average contributors to the bottom line. None of this is encouraging to industry.

The US Market Share is Stagnant or Shrinking. Will the Industry be Able to Prioritize Ex-US Territories?

The pharmaceutical industry is going to come under even greater financial pressure in the years to come. In 2002, the US accounted for about 50% of all pharmaceutical dollars (not counting generics) followed by Europe with 20%, Japan with 10% and the rest of the world at 20%. This position has held constant over several decades. This is interesting since Europe, for example, has a population 30% larger than that of the US. Europe may use fewer drugs, but their cost is also about 30% less than the cost of the same drugs in the US. One big reason for this is the fact that the US is one of the few countries that does not have a national negotiation for drug prices. In spite of this, with the economic downturn of 2008–2009, the pharmaceutical market in the US is projected to account for less than 40% of the world market. The growth area is now in the developing world where emerging markets are expected to account for 14–15% of the global market in 2009 according to IMS Health. The loss of market share in the US is related to the reluctance (or inability) of consumers to spend given their current financial constraints, but also to developments in the pharmaceutical marketplace itself. There are more and more large health cooperatives and managed care organizations that are negotiating lower prices with industry. Because these cooperatives can command large sales volumes, industry is forced to negotiate. This situation will only progress as the US moves towards an inevitable reform of healthcare overall. The pharmaceutical industry will have to adapt to these changes by expanding into emerging markets and by adapting to new standards appearing in Europe and in the US. We will discuss this further in Chapter 7. But my point here is that this pressure may in turn increase the pressure on the industry to turn away from less profitable drugs like antibiotics.

As far as the US regulatory climate is concerned, the decreasing influence of the US marketplace may tempt companies to deprioritize the US at some point. Thus, as I noted in Chapter 4, our antibiotics regulators will have truly succeeded in regulating themselves out of business and the American public out of new antibiotics.

By the way, one interpretation of these market figures is that the US is actually subsidizing the discovery and development of drugs for the rest of the world. Most large companies pour somewhere between 10 and 20% of their revenues back into research and development. According to the market numbers I noted where the US has accounted for 40–50% of the market globally, we also contribute 40–50% of the research and development dollars to the pharmaceutical industry. We therefore subsidize the discovery and development of new drugs for the rest of the world. The logical conclusion of this reasoning is that as the pharmaceutical market shrinks globally, or at least grows less quickly, there will be fewer dollars going into the discovery and development of new drugs. Some argue that Europe and Asia will pick up the slack left by the US, but with their strict price controls and national negoatiations for drugs and drug pricing, I can't imagine that this will occur. There will therefore be fewer new drugs. I'm not sure whether this is a good thing or not, but I predict that it will become reality as the US shrinks its overall market share and the market itself grows more slowly or even shrinks.

Research and Development Costs Increase, but Approvals Are Down

The other major pressure on pharmaceutical companies is the ever-increasing expense of research and development. Companies have been spending more and more on research and development both in terms of absolute dollars and as a percentage of revenues. This increase is illustrated in the illustration below that comes from the web site of PhRMA. In 2007 PhRMA member companies spent a massive $103 billion on research and development. It is estimated that, partly because of the cost of drug failures, each new drug approved has an overall cost to the company of 800 million to $1.2 billion. The money poured into R&D by large PhRMA continues to increase as shown below (Fig. 6.3). Also shown below (Fig. 6.4) is the fact that the number of new molecules (NMEs) that are actually approved has been decreasing with time in spite of a continued increase in the number of molecules entering the early stages of clinical development (INDs). That is, just looking at brand new molecules and not new uses of old ones, the numbers that enter the early stages of clinical development have been increasing. But, the proportion of those that make it all the way to market have been decreasing. So the industry is achieving less and less bang for more and more dollars. This squeezes their budgets forcing

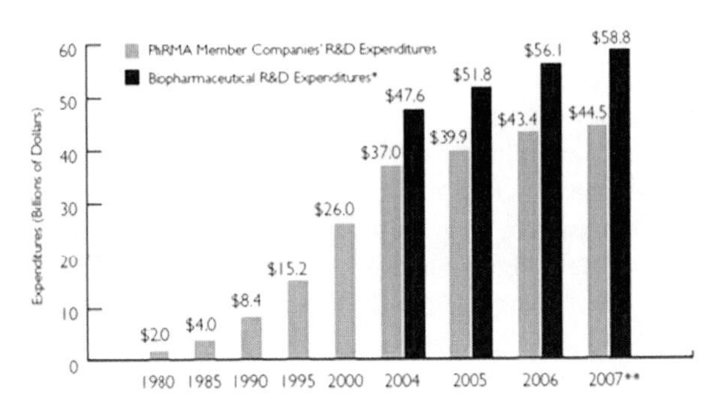

Fig. 6.3 Large pharmaceutical companies expenditures on research and development over time. Taken from the website of the Pharmaceutical Research and Manufacturers' web site with permission

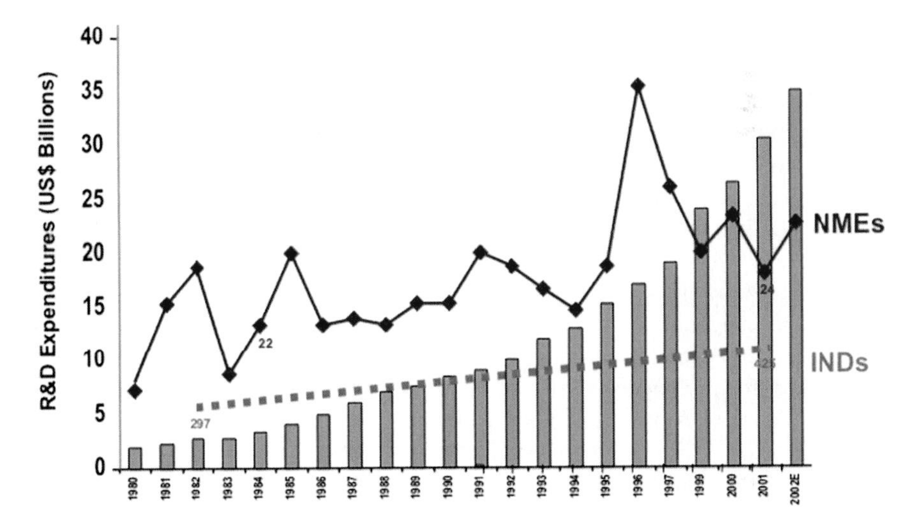

Fig. 6.4 The production of new drugs entering clinical trials (INDs) is not following the increasing investment in pharmaceutical R&D. Figure kindly provided by Dr. Steven Projan

them to prioritize among their many projects to favor those projects that have the most promise in terms of both success and return on investment. That usually means that antibiotics get short shrift.

Some Good News for Everyone

In spite of all this doom and gloom, there is a recent success story that should provide lessons for everyone, including large pharma. Linezolid is an antibiotic with a long and tortuous history. Back in the late 1970s, DuPont simply tested its chemical library of compounds for those that would inhibit the growth of bacteria. They identified a couple of molecules with a similar chemical structure that seemed to work both in the test tube and in animal models of infection. These molecules, though, were toxic. In the 1980s, Pharmacia-Upjohn (now part of Pfizer) discovered a molecule in the same chemical series, the oxazolidinones, which was active in test tubes and worked when given either intravenously or by the oral route in animal models. It was less toxic than the earlier molecules. The compound, called linezolid (Zyvox) was active against all the resistant strains of staph and against the highly resistant enterococci that were becoming so problematic in US hospitals at the time. Pharmacia tested the antibiotic in patients with infections due to these highly resistant pathogens and it worked. Not only that, but the blood levels of the antibiotic were the same whether it was given intravenously or orally. This was very unusual and remains a very important advantage. It means that a patient can be started on the IV form when they are very sick in the hospital and later, when they are starting to get better, they can simply go over to the oral form of the drug. Pharmacia realized that this might even allow patients to be discharged earlier from hospitals and they designed the clinical trials to find out whether that was true or not. It was. Knowing that this molecule would be sold, at least initially, in the hospital, and that it would save hospitals money by allowing early discharge, they made the selling price very high. The high price helped make up for the very low prescription volume, since only those patients with demonstrated or suspected resistant infections would be treated. In the test tube and in animals, it took a very long time for resistance to develop if it happened at all. Based on what we now understand about the way linezolid works, this makes good scientific sense. The one major drawback to the molecule is that it still retains some of the toxicity that was seen by DuPont at the beginning. It tends to cause some suppression of the bone marrow when given for prolonged periods of time – say more than 10–14 days. Because it is used for serious infections with resistant bacteria, patients and physicians are willing to deal with linezolid's toxicity. Linezolid became the first oxazolidinone to be approved and marketed in 2000. It is now selling around $1B. There are at least two new versions of oxazolidinones in late stage clinical development in biotech companies.

Linezolid provides several valuable lessons for us all. First, it demonstrates that occasionally, using very old methods of antibiotic discovery, just looking for things that kill bacteria, you can come up with a winner. Second, by showing that your molecule is cost effective, you can charge a price that is justified by the money saved. This means that even with only a relatively few prescriptions, you can still have a good return on your investment. This is one of the few examples of the use of pharmacoeconomics to allow an antibiotic to succeed in a small niche market. For any company thinking of antibiotics, linezolid should be a bright light showing at least one promising way forward. Cubist and daptomycin successfully followed this path later.

Another boon from linezolid is that they have created a niche market that can now be exploited by others. A number of companies are now targeting oral drugs for the treatment of MRSA infections. They all hope that their drugs will be as effective as linezolid, but will be less toxic such that they can gain linezolid market share. Unfortunately, most of these companies are small biotech ventures and as such, their risk is higher.

Antibiotic R&D in Biotech Is Stymied by Late Stage Clinical Trial Costs

With the abandonment of antibiotic research by many large pharmaceutical companies, there has been a flurry of divestments of antibiotics in various stages of development to small pharmaceutical companies. This has generally been to the benefit of biotech companies. Only a few have ultimately succeeded, though. At least one example of this probably has the large pharmaceutical company asking themselves what they did.

Back in the 1990s, Lilly was testing an antibiotic called daptomycin. It was interesting because it was active against the vancomycin-resistant enterococci that were just emerging at the time (see Chapter 3) and was also active against MRSA, offering an option beyond vancomycin for the treatment of these staphylococcal strains. The drug was causing muscle toxicity in the clinic. At the same time, they were testing it in one of the most difficult infections you can treat, heart valve infection. They decided that the dose required for success in heart valve infections was too high for the level of muscle toxicity they were seeing and they halted development. Shortly after that decision, they changed strategy entirely and abandoned antibiotic research altogether. Lilly then set out to divest daptomycin. At the same time, Cubist, a biotech company in Boston, was selling technology on genomics to large pharmaceutical companies. But Cubist saw the light, or should I say dark, at the end of that tunnel. By then, investors were turning away from financing pure technology since they discovered that the runway to profitability for the company was too long. Realizing this, Cubist decided to try and bring in a drug candidate that would allow them to attract new investor money. Cubist looked carefully at daptomycin. They theorized based on Lilly's own data, that they would be able to treat infections and avoid toxicity by altering the dosing regimen that Lilly had used. Frank Tally and others at Cubist showed that the antibacterial activity was best with infrequent dosing while muscle toxicity was worse with frequent dosing. So – they licensed the drug and started early stage trials using less frequent but higher doses than Lilly had used. It worked. They saw very little muscle toxicity and the antibiotic worked against serious skin infections. Cubist then had to figure out how it was going to fund their late stage pivotal clinical trials that are so expensive. It is hard to know what happened at Cubist during those years, but what they ended up doing was going to the public markets to fund their late stage trials. This was a very risky, gutsy, and ultimately successful move. They remain one of the very few biotech companies that have been able to fund late stage trials at the $100 million or more level without partnering with or being purchased by a large pharmaceutical company. Daptomycin was a commercial success in the US beyond the expectations of

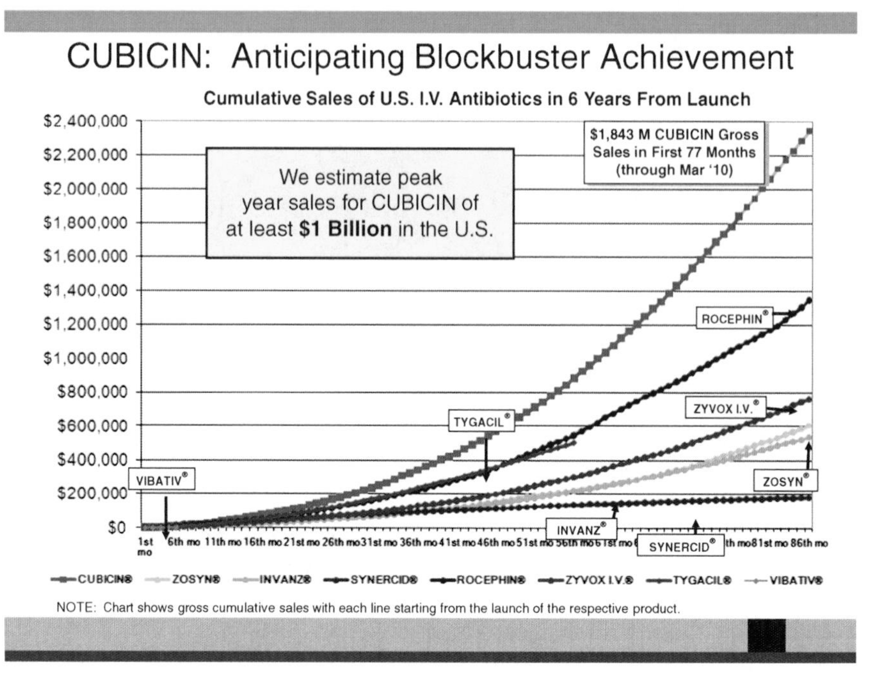

Fig. 6.5 Sales of Cubicin (daptomycin) and other antibiotics with time after launch. Taken from the Cubist website with permission

most of us pundits, and today brings in well over $400 million dollars in revenue. The chart (Fig. 6.5), taken from Cubist's corporate presentation available on their website, compares daptomycin (Cubicin) with other antibiotics. In the chart, Zyvox, Zosyn and Rocephin all achieved $1B or more peak year sales.

For any company, large or small, this would be considered a blockbuster. I wonder what Lilly is thinking now. Of course, they do get royalties based on daptomycin sales.

The chart from Cubist illustrates another good point about antibiotics. It takes years, sometimes 10 years, for them to achieve their peak sales. For pharmaceutical companies, this means that it may take a long time for them to make up their development costs. There are several explanations for this. Unless a new antibiotic immediately addresses a severe and widespread problem of antibiotic resistance, there is little motivation to use the drug immediately. A new drug will certainly be more expensive than older, generic drugs. Knowing that resistance to the new agent will eventually emerge, physicians try to carefully manage their use of the new antibiotic to avoid rapid emergence of resistance. This is good public health policy, but it slows sales. So far, no recently launched antibiotic has come out at a time where resistance was such a problem that most physicians were obligated to turn to the new drug as a drug of first choice. This is a great thing because it means that we have always had an armamentarium of antibiotics active against emerging resistant strains in reserve. It also means that something was there for those patients that needed it. At the same time, the resulting slow sales curves have further discouraged companies from antibiotic research. As Dr. Steven Projan of Novartis has pointed out, antibiotics is the only therapeutic area where a promising, new, active and safe compound is put on the shelf after approval.

Biotech Is Still a High-Risk Proposition

For every biotech success story like Cubist, there are many less happy stories. Some experts estimate that only 10% of biotech companies will succeed. One example of the other 90% is Replidyne, of Louisville, Colorado. Replidyne started with technologies for finding new antibiotics, but then realized that it too needed a near term product to attract investor capital. They licensed an antibiotic from Glaxo Smith Kline that could only be developed as a topical product. This was not sufficient to attract investor dollars, so they then licensed in an oral antibiotic called faropenem from a Japanese pharmaceutical company, Daiichi-Suntory, now Asubio. Faropenem was like an oral version of the very active carbapenems like imipenem. It had already been tested in late stage trials by Bayer (who had licensed it previously from the Daiichi). (In fact, when I first joined Wyeth, we were just giving faropenem back to Daiichi. It is such a small world!) When Bayer abandoned antibiotics, they returned the license to the Daiichi and gave them all their data and reports. The dossier was never submitted to the FDA for approval. Replidyne carried out a reanalysis of the trial data, and, in 2005, they submitted an NDA (New Drug Application – a request for approval) to the FDA for sinusitis, bronchitis, pneumonia and skin infections including data from 11 pivotal clinical trials. The FDA rejected their application for all indications in 2007. The FDA said that for sinusitis and bronchitis they now required placebo-controlled trials and the trials submitted by Replidyne (actually performed by Bayer) were of a different design (see Chapter 4). Neither Replidyne nor Bayer had any idea that placebo-controlled trials would be required at the time. The FDA said that the data in the pneumonia package did not have enough microbiology data, and two additional studies would be required for approval in that indication. At the time, Replidyne was a publicly traded company. They did not have to finance the pivotal trials they had submitted since they had largely been carried out prior to their licensing the compound. They did have to pay for the license from Asubio, though. Replidyne had lined up Forest Pharmaceuticals in the US as a partner to sell faropenem. With the FDA response, Replidyne attempted to carry out a placebo-controlled trial of faropenem in bronchitis. They failed since physicians very simply would not enroll patients in a trial where they might get no therapy. Forest dissolved its partnership with Replidyne. Eventually, Replidyne went belly up and sold their public listing to a medical device company. Everyone working on antibiotics (essentially the entire company) was laid off. Unfortunately, this has been the more typical story for biotech in general including those companies trying to get into the antibiotics game.

The other problem in biotech working on antibiotics, just as in large pharmaceutical companies, is that of innovation. No one (almost) seems able to come up with novel therapies. If we look at the chart below, we can see that all but three of the antibiotics in late stage development or just finishing development from biotech had their origins in large pharmaceutical companies, just like daptomycin from Lilly was sold to Cubist who then developed it. In Table 6.5, the antibiotics shown in italics were actually discovered in the biotech company that developed it. All the others were discovered in large pharmaceutical companies and licensed to the biotech companies. This indicates that (1) large pharmaceutical companies do not see value in products that biotech do value and (2) that innovation in biotech has been poor in that very few products are actually discovered there.

Table 6.5 Antimicrobial compounds in development by biotech

Compound	Manufacturer	Company of origin	Status
Dalbavancin	Pfizer from Vicuron	Merrell Marion Dow	Unknown
Iclaprim	Arpida	Roche	Failed in US
Oritavancin	Targanta	Lilly	Failed in US
Telavancin	*Theravance*	*Theravance*	*Approved*
Ceftibiprole	Basilea-Johnson&Johnson (J&J)	Roche	Delayed in US
Cethromycin	Advanced Life Sciences	Abbott	Failed in US
Doripenem	Peninsula J&J	Shionogi	Approved
*EDP-420**	*Enanta*	*Enanta*	*Ph. II*
Faropenem	Replidyne	Daiichi Suntory	Failed in US
Ceftaroline	Cerexa/Forest	Takeda	Submitted for approval
NXL-104	Novexel	Aventis	Phase II
NXL-103	Novexel	Aventis	Phase II
PTK-0796	*Paratek – Novartis*	*Paratek*	*Phase III*
Torezolid	Trius	Dong-A	Ph. II
Radezolid	*Rib-X*	*Rib-X*	*Ph II*

*EDP-420 is being developed for community-acquired pneumonia as an oral antibiotic. As such, the trial design will be virtually impossible to carry out. See Chapter 4.

On a more optimistic note, several biotechs are nearing Ph. II development with compounds actually innovated within their biotechs. *Nabriva* has a series of pleuromutilins ready for Ph. II trials. They are active against Gram-positive skin pathogens and both typical and atypical respiratory pathogens. *Tetraphase* has a lead molecule with broad-spectrum activity that comes out of a novel and exciting method of synthesis of tetracyclines also poised for Ph. II trials. *Achaogen* has a lead aminoglycoside that they discovered which is ready for Ph. II. Maybe innovation in biotech is going to actually produce some new antibiotics active against resistant pathogens. This would be welcome news for us all.

What about the future for antibiotics as a product for biotech companies? Not much has changed. The late stage trials, because of the ever-increasing stringency required by the FDA are becoming more and more expensive putting them farther out of the reach of biotech. Certainly, biotech companies with no source of revenue (most of them), and with only private financing (again, most of them), are not usually able to pay for such trials. They therefore either have to go to the public marketplace for such funding (the public markets have been closed to new entries for most of 2008 and 2009) or they have to sell themselves or their product to a large pharmaceutical company. But, as we've seen above, there are fewer and fewer large pharmaceutical companies extant and of those that exist, fewer are interested in antibiotics. Of those seven companies no longer pursuing antibiotics research, only one, J&J, actively acquires antibiotics from biotech. So I would say that the future of biotech developing antibiotics is at best shaky unless something changes on the funding side or the cost side. Given that 90% of biotech will fail anyway, at least according to some experts, I'm not sure that antibiotics are any more risky than any other therapeutic area in this regard.

Chapter 7
Modest Proposals

How are we going to get out of this mess? First, I'm not sure that we will. To extricate ourselves from a problem of this magnitude with this degree of complexity will not be easy. But I can suggest some directions that might get us down the right path.

The Role of Government

We need an overall, coordinated attack on antibiotic resistance as it currently stands. Only by succeeding in stemming the rise of resistance will we maintain the effectiveness of the antibiotics we have currently. And we will need them since the antibiotic pipeline is still relatively dry.

Government Loves a Task Force

In 1999, the department of Health and Human Services formed the Interagency Task Force on Antimicrobial Resistance. The basic premise was to get various government agencies form a common plan of attack for the problem, to communicate with each other and to develop cross-agency plans and projects. Aside from those agencies like the CDC, NIH and FDA, the Task Force ultimately included the Agency for Healthcare Research and Quality (AHRQ), Centers for Medicare Medicaid Services (CMS), the Health Resources and Services Administration (HRSA), the Department of Agriculture (USDA), the Department of Defense (DoD), the Department of Veterans Affairs (VA), and the Environmental Protection Agency (EPA). In this way, all aspects of the problem of resistance, antibiotic overuse, infection control, surveillance for resistance, antibiotic use in animals and crops, environmental contamination with antibiotics and our suffering antibiotic pipeline could be addressed in a coordinated way. Great idea! But what makes any of us think that the rest of government is better able to connect the dots than our various intelligence agencies?

D.M. Shlaes, *Antibiotics*, DOI 10.1007/978-90-481-9057-7_7,

The plan is available online through the CDC website (www.cdc.gov). Some agencies, especially the CDC, have done a good job in educational outreach of antibiotic resistance and have carried out some important surveillance under the auspices of the Task Force. The NIH, in 2006, established the Drug Discovery and Mechanisms of Antimicrobial Resistance (DDR) to provide appropriate expertise to review research grants on resistance. (Of course it took them 20 years to put this in place since it was first recommended by another task force back in 1986). Even though this year only 1 in 17 grants will be funded, NIH has succeeded in reversing a trend going back to the 1950s of not funding antibiotic research. The Centers for Medicare and Medicaid Services, as part of their quality initiative for hospital reimbursement has identified certain hospital acquired infections for which it will no longer pay. This one initiative may have considerable impact even though it is a controversial one. The FDA, as far as I can tell, has slid backwards in their goals of stimulating antibiotic discovery and development. Large pharmaceutical companies have continued to leave the field in the last decade and essentially none of those who halted antibiotic research have come back to it.

If I had to give the Interagency Task Force a grade after a decade of work it would be D. Why are they doing so poorly? There are two major reasons and you can guess what they are. Money is number one. These agencies all compete for funding within the federal budget. Those within HHS compete for HHS monies. The goals of the Task Force require funding which has never been forthcoming in any kind of systematic or dedicated way. Yet we lose more Americans every year to antibiotic resistance than to terrorism, automobile accidents, and war. We lost 58,236 Americans in the entire Vietnam War, but we lost 63,000 per year to antibiotic resistance just in our hospitals alone! That translates to over 630,000 Americans lost to resistance since the inauguration of the Task Force. Where are our funding priorities?

The other issue is that these agencies do not communicate well. The actions already undertaken by the various agencies were the result of efforts within each of them individually. Where is the coordinated attack? Where are the cross-agency goals and projects? This is related to the first problem – they compete for funding. They are all large bureaucracies that do not change their behavior quickly.

In my view, we need a centralized and independent authority with expertise in multiple facets of resistance and a reasonable budget. This budget can come in part from all the agencies already part of the Task Force. Let the agencies have some of their budget monies dangled in front of them that they would only be able to access with practical, important and coordinated plans to deal with multiple facets of the resistance problem. Let them compete for grant monies for a change.

FYI – We have just agreed to establish the EU/US Transatlantic Taskforce on Antibiotic Resistance, focused on appropriate use of antibiotics and strategies for improving the pipeline of new antibiotics. Lets hope it works better than the Interagency Task Force that has been in place for the last 10 years.

Government Can Play Several Important Roles

First, government can provide important basic science research and training that would both support our efforts to improve our effectiveness in the battle against resistance and help industry to find new antibiotics by creating a qualified workforce and with advances in science. Government should take a lead role in the fight against resistance by coordinating efforts across the CDC, the FDA and the NIH all of which are part of the Department of Health and Human Services. Government can provide appropriate incentives to industry to discover and develop antibiotics for resistant infections. These actions could be seen as part of the public health responsibility of government. The FDA can make a huge difference by providing a more balanced approach to clinical trial design requirements. Finally, the pharmaceutical industry itself can be part of the solution in several ways.

To battle resistance, we need a multi-front war. Clinicians, epidemiologists, researchers and regulators all need to be on the front lines. Unfortunately, this cat is already out of the bag and it is unlikely that we will be able to stem the current tide of resistant staph and Gram negatives – the so called ESKAPE organisms of the Infectious Diseases Society. Nevertheless, we must build a capability to protect the next generations of antibiotics that are ultimately developed. To accomplish this, there are many scientific questions that remain unanswered for lack of available funding. Does longer therapy engender more resistance or does shorter therapy? Do we really need to treat pneumonia for 7–10 days? Do we need to treat urinary tract infections for 1, 3, 5 days or longer? When the FDA approves an antibiotic, it approves its use during the length of therapy that was used to study the antibiotic in its late stage clinical trials. But there has rarely been the kind of systematic study of length of therapy that is needed to determine what the correct length should be. Are the current guidelines on length of therapy for various infections justified by the data? The answer to the last question is, in general, no. The NIH is now supporting trials to examine this question in a few infections, but this is a very open area for research.

Another important question has been raised by the epidemic of MRSA that is sweeping our nation and many others worldwide right now. It turns out that many patients can carry MRSA on their skin and in their nose for prolonged periods of time. If they become carriers (colonized), they become a reservoir of MRSA for others including family members workmates and, in the hospital, health care workers and other patients. Some countries have strict measures in place in their hospitals to prevent the spread of MRSA. These measures start with obtaining swabs of skin and nose to look for MRSA in everyone who is admitted, or, sometimes, screening only patients that are at high risk of carrying MRSA. The strict measures then call for placing those patients with positive cultures in isolation. In this case, they are placed in a private room. Visitors and health care professionals entering the room are usually required to wear gowns and gloves and sometimes masks as well. This restriction carries the risk that the patients will get less attention from health care professionals and it certainly is an inconvenience for visitors. Some countries or institutions also require that the health care providers attempt to rid the patient of their MRSA carriage state with disinfectant baths and topical antibiotics. These countries and institutions believe that this has helped stop the spread of MRSA. But has it? There are precious few data that clearly support these policies and whole countries and individual hospitals are on their own here. This is an area that cries out for government-sponsored research to answer this question once and for all. It is not inexpensive to isolate patients or to try and decolonize them. It is also not so pleasant for the patient who is colonized, their families and their physicians. As I noted, there are even data suggesting that patients in isolation get less attention from health care providers because of the isolation procedures that are used (wearing gowns, gloves, keeping the door to the room closed, etc.). The CDC has refused to recommend these measures routinely in US hospitals for lack of clear data supporting their use. Again, both the NIH and the CDC can take lead roles here.

The Infectious Diseases Society and the Society for Health-Care Epidemiology have issued broad guidelines on the prevention of antibiotic resistance in hospitals. These guidelines relate to antibiotic use policies in hospitals, infection control policies including whether or not patients should be screened for MRSA, environmental cleaning policies and even administrative procedures. Because data supporting their guidelines are often lacking, they have also identified key areas of research to better inform those responsible. These research efforts need support from many different stakeholders ranging from the CDC to hospital administrators to those who approve hospitals for reimbursement by Medicare and others. Again, many of the recommendations in these guidelines need additional justification that can only come from more research. Again, the NIH and CDC can take an important lead here.

The use of human antibiotics in animals, either for growth promotion or for therapy should be addressed immediately as should the use of antibiotics as crop pesticides. Here the FDA should be taking a lead role.

In fact, in the 1990s, the government formed a task force uniting representatives from NIH, FDA and CDC to provide an action plan to combat resistance (I was on their initial advisory board). The scope of their brief was very broad. They have issued several reports where goals for each agency were constructed to fight resistance and progress on achieving these goals has been tracked. In spite of much talk and some notable accomplishments, in my view, we could do much more. An update from the Task Force was due in 2009, but – surprise! – has not appeared. The goals enumerated by the interagency task force have been allowed to languish in a chronic lack of funding. In addition, the fact is that these agencies do not communicate very well with each other. They do compete for federal dollars after all. Nevertheless, they are all working under the Secretary for Health and Human Services. The Secretary should be lobbying for appropriate funding and should be holding the collective feet of the various agencies involved to the resistance fire.

The Transatlantic Taskforce on Antimicrobial Resistance

At the EU-US Summit on the 3rd of November, 2009, it was agreed that a transatlantic taskforce on urgent antimicrobial resistance issues be established. The task force is to focus on appropriate therapeutic use of antimicrobial drugs in the medical and veterinary communities, prevention of both healthcare- and community-associated drug-resistant infections, and strategies for improving the pipeline of new antimicrobial drugs, which could be better addressed by intensified cooperation between the United States and Europe. The infectious Diseases Society of America has set a goal of 10 new antibiotics approved by 2020. They have also made a number of specific recommendations for the establishment of a number of working groups for the taskforce. I worry that this will be a lot of bureaucracy and academic discussions with little result. But I am hopeful, and obviously I support and commend the efforts of Sweden, Europe and the Obama administration. These events suggest that governments are beginning to see the seriousness of the problem we are facing. Will they act?

Government-Sponsored Research and Research Tools Are Required

Ignoring the problem of the FDA for a moment, one might ask whether there is an appropriate role for government outside the obvious role government plays in the battle against resistance and in the support of antibiotic research. Many might reply that if industry will not discover and develop the antibiotics that we need, this should be a role for government. Antibiotic resistance is, after all, a public health issue. Although this is true, the government has not yet shown it is capable of discovering and developing drugs on its own. There are probably many reasons for this. There is a lack of expertise within government as well as a lack of appropriate infrastructure. When I just consider the problem of drug manufacturing alone, I am overwhelmed by the idea of the government managing all this. With the appropriate will and funding, I am convinced that such an effort ultimately would be feasible. But I also think that the years it would take to establish the appropriate expertise and infrastructure plus the inevitable slowness of the government bureaucracy would make this choice the least efficient one.

The expertise in antibiotic discovery and development still resides within the pharmaceutical industry. The best approach here would be to find a way to harness what already exists and turn it to the public health need for treatment of antibiotic resistant infections that will surely worsen with time. I think the most important contributions government can make at this point in time are: (1) Providing appropriate incentives to industry, large and small, for the introduction of new antibiotics which provide activity for resistant bacteria as designated by the Infectious Diseases Society of America's priority list, the ESKAPE organisms discussed in Chapter 3. (2) Further the government should increase funding for discovery and early development of new antibiotics within academia. Since the science is hard, having more hands working on it makes sense. (3) The NIH should immediately address those clinical trials which industry is unlikely to carry out. The first priority would be placebo-controlled trials for exacerbations of chronic bronchitis. For a full explanation of this issue, see Chapter 4. Other trials that are needed include studies on infections of heart valves and bones and joints. Since these infections are rare, the NIH could establish consortia of centers with access to large patient populations with these difficult infections. This would make NIH sponsored trials easier to perform and might even tempt industry back into this difficult area.

The NIH could also be very helpful in doing appropriate trials in the pediatric population. One example of this is the ongoing Finnish placebo controlled trial in children with well-documented ear infections.

Finally, the NIH could take a leadership role in providing training in scientific disciplines that have disappeared with the decrease in funding for research in what used to be called bacteriology. They have taken steps in this direction in recent years, but this effort could be pushed harder. Trained scientists are required if industry is going to continue to try and find and market new antibiotics, and will also be required to support the government funded research we discussed above.

Industry Incentives Will Probably Also Be Required. Europe Is Leading the Way in Thinking Here

An incentive for industry is a very touchy subject. Incentives will mean that taxpayers will end up paying an industry they dislike (hate?) to provide therapies that the industry views as not sufficiently profitable for their shareholders. What incentives will work for industry? The Infectious Diseases Society of America has made a series of recommendations in this regard. One major recommendation they made in their position paper of 2004 was to provide a wild card patent exclusivity to companies who successfully brought an antibiotic with activity against key resistant pathogens to the market. What this would mean would be that the company would be allowed from 6 months to 2 years of additional market exclusivity in the US for a drug of their choice within their portfolio. For example, if a company were to bring gorillacillin, an (imaginary at this point) antibiotic with activity against highly resistant Gram negatives like Pseudomonas to market, in addition to the money they might make on that antibiotic (probably not a lot of money) they could continue to sell their blockbuster anti-cholesterol drug for an extra 6 months to 2 years before generics would be allowed onto that market place. That means an extra 6 months to 2 years of high costs for consumers, insurance companies and Medicare Plan D for the cholesterol drug to help provide profits to the company for bringing a needed antibiotic forward for patients and physicians. This incentive, I know, would work. Not only that, but an analysis by Brad Spellberg and the Infectious Diseases Society of America has shown that that the wild card exclusivity extension would be cost effective! The problem is that the generic companies hate it, consumers hate it and congress hates it, so it has been DOA in congress even before the Infectious

Diseases Society proposed it. Other incentives proposed have included extensions of the patent exclusivity on the antibiotic itself. This is not too interesting to industry since their sales of antibiotics are not so profitable to begin with.

Another tactic that has been proposed by Dr. Louis Rice of Cleveland is to provide disincentives for companies that abandon antibiotics research. In his model, tax incentives like deductions for research and development would be decreased for companies that no longer work on antibiotics while they would stay in place for those that continue. I personally feel that this plan would be very difficult to implement, but it seems interesting enough to discuss further.

Europe has made a number of very interesting proposals. I must admit that I am astonished by the enlightened approach coming from Europe. First, it is ironic that the leader in the effort to find a solution to the perfect storm surrounding antibiotics has been Sweden. Sweden has some of the lowest rates of antibiotic resistance in the world. Second, I have a jaundiced view of Europe having lived there for several years. I view Europe as a huge bureaucracy uniting (poorly) 27 countries with different languages, sometimes even with different alphabets, disparate cultures and completely different attitudes towards many aspects of life. And yet, some of the best and most profound thinking on the antibiotics problem is coming from Europe. And, if Europe is able to implement its key proposals, they will lead the way forward for all of us.

As discussed in Chapter 5, the basic premise is to use a combined "push-pull" mechanism. In one embodiment of such an incentive, Europe would essentially take on some proportion of the development costs for a compound thus reducing or eliminating the risk for the company depending on the amount provided. This, in their model, is equivalent to purchasing an option to buy quantities of the antibiotic should it be approved by an appropriate regulatory body and should it meet preset criteria such as activity against resistant pathogens (which would in turn require more precise definition). This would be the "push" part of the mechanism. If the antibiotic were to be approved, Europe would then be obligated to purchase a set amount, I presume in terms of courses of therapy, at some set price. This would be the "pull" portion. The attractiveness of this model to industry would largely depend on the size of both the "push" and the "pull."

Many companies use a calculation of the presumed Net Present Value (NPV) of a product to decide whether or not it is worth pursuing. It is because of this calculation that many companies have abandoned antibiotics since the ultimate sales do not provide much value above and beyond the costs of development. In the calculation of NPV for a product, upfront expenses like those for late stage clinical development and the early marketing expense for launching the product count heavily. At the same time, later earnings are discounted for inflation. So companies are always looking for products that provide for high, early sales – "a steep revenue curve." The European approach is attractive when viewed this way. The "option" on the product will relieve a portion (hopefully a large portion) of the upfront expenses for late stage clinical development. The "pull" would bolster the early portion of the sales curve. Thus the push-pull approach, if both components were large enough, would have a substantial effect on the overall NPV for the product.

The FDA (and EMEA)

The FDA and EMEA Both Must Adapt a More Balanced Approach to Antibiotics. Required Trial Designs Must Be Feasible

As I considered my remarks for the FDA advisory committee reconsidering guidelines for clinical trial design in community acquired pneumonia, I asked myself, "Why would the FDA require clinical trials that are infeasible?" Perhaps they believe that good science requires such designs – but that gets us nowhere. Maybe they simply have decided that we won't need new antibiotics for respiratory tract infections for at least the next decade. In that case – message received. Another thought I had was that they simply don't realize that their proposed trials are infeasible. If that is the case, they have clearly lost touch with reality and have become so divorced from the industry that they regulate that they are no longer capable of drafting reasonable guidance. Finally, the most distasteful possibility is that politics is playing a major role in the decision making at the FDA. In my view, politics has no place in our discourse.

One approach to regain balance in the FDA's approach for antibiotics is for them to reconnect with industry. The FDA must come to grips with the fact that we need a vibrant pharmaceutical industry to bring new antibiotics to the marketplace and that requiring infeasible trials will simply drive the industry away from antibiotics and into more profitable ventures. The industry is, after all, not the enemy here. We need to work together. That is, we all have to be able to play in the same sandbox without throwing all the toys out.

Do We Want New Antibiotics for Mild Infections? Is Bacterial Bronchitis in the Setting of Chronic Lung Disease a Mild Infection?

The FDA and Europe could make an enormous difference immediately. First, for mild, acute infections like otitis, sinusitis and bronchitis, they need to reconsider their entire approach. For example, for otitis, where most authorities agree that expectant therapy is a reasonable approach, placebo-controlled trials remain difficult to accomplish, especially in the US. Guidelines from the American Society of Pediatrics suggest that patients with severe symptoms, those age 2 or less and those where the diagnosis is certain that it is otitis media, be treated with antibiotics immediately. This leaves older children, those with milder disease and those where the diagnosis is less certain (the majority of patients) available for expectant therapy. Nevertheless, recent surveys have shown that only 15% of children in the US are treated expectantly. The most common reason is parental concern (85% of parents) about not using antibiotics. One solution is based on a recent study reviewing many placebo-controlled trials in otitis. The data indicate that between days 2 and 4, 25% fewer children on antibiotics will have continued symptoms compared to patients on placebo. If this were true, one could design a comparative trial where symptoms on day 4 were the endpoint. Comparing two antibiotics with a margin of 10% (at worst, the new antibiotic could be no more that 10% worse than the comparator treatment) should be acceptable. The FDA may argue that since the data does not come from a single large placebo controlled trial it is invalid. Of course that single large placebo controlled trial may not be feasible. If we ever want to have new antibiotics for ear infections, the regulatory agencies must find another way forward.

A similar approach could probably be taken with sinusitis. In one study, only 11% of 44 illnesses in placebo recipients had completely resolved by day 10 compared with a complete resolution rate of 38.5% (32 of 83) among patients treated with antibiotics. By day 10, a total of 25 patients (57%) in the placebo group had complete resolution or were much better, compared with 71 (85.5%) of those receiving antibiotics. These data would suggest a treatment effect of about 30% (that is 30% more antibiotic treated patients were symptom free than placebo treated patients) for an endpoint of complete resolution by days 5–10. Again, a comparison of two antibiotics for this endpoint with a margin of 10% should be very acceptable. The FDA and EMEA could again make the same counter argument here as I noted above for otitis. My response is also the same. They have to find a way forward other than placebo-controlled trials if we ever want to have new antibiotics for sinusitis.

For severe exacerbations of chronic bronchitis, the regulatory agencies simply have to look at the Cochrane report and clinical practice guidelines, and then back off their current requirement for placebo-controlled trials. Administration of a placebo to these patients would probably not be ethical, clearly contradicts current practice guidelines, and may put patients' lives at risk. According to the Cochrane database, antibiotic treatment of COPD exacerbations with increased cough and sputum purulence reduces the risk of short-term mortality (as in DEATH!) by 77%, decreases the risk of treatment failure by 53% and the risk of sputum purulence by 44%. Given these impressive numbers, we could easily justify a return to our previous method of carrying out comparative trials of antibiotics with a statistical margin of 10% and be more than satisfied.

To be fair, the proposed changes to the European guidelines (not yet approved) indicate more flexibility in their approach to these so-called mild bacterial infections. But this approach leaves everything uncertain. Maybe if Europe took a courageous stand here the FDA would follow.

The FDA Must Modify Its Guidelines for Clinical Trials in Community Acquired Bacterial Pneumonia

As I noted earlier, the FDA's new guidelines for the conduct of clinical trials in community-acquired pneumonia require an extraordinary number of patients to test treatment in outpatients with this disease. This is for two reasons. First, the FDA requires that the physician actually have identified the bacterium causing the pneumonia and they require that it only be certain bacteria where the FDA is sure that antibiotic treatment works. From the point of view of actually carrying out a trial, this means that about 2500 patients might have to be enrolled per trial over two trials (5000 patients total). Under the previous system where a microbiological diagnosis was not required, we enrolled about 700 patients per trial or 1400 patients total over two trials. Why are the trials requiring a microbiological diagnosis so large? First, you can't obtain a specimen from all patients from which to isolate or detect the bacterial pathogen. Second, you don't always isolate the bacterial pathogen even when you do obtain a specimen for culture. Third, even when all the above are successful, not every patient with pneumonia has one of the bacterial pathogens causing the infection that the FDA will accept.

The FDA also requires that no prior antibiotic therapy be given. These patients are hard to find and are becoming more and more rare. Why is it so hard to find patients with pneumonia that have not already been treated with an antibiotic? First, people have telephones. They call their physician who will sometimes prescribe an antibiotic over the phone without realizing that the patient has pneumonia. Second, to maintain their accreditation, hospitals have to comply with a specific measure of quality regarding patients with pneumonia. Those patients must receive their first dose of antibiotic within 6 h of presentation to the hospital. Therefore, there simply is not time to enroll them in a trial prior to that first dose.

Putting these two requirements (a microbiological diagnosis and no prior antibiotics) together will make conducting these trials exceedingly expensive, slow and almost impossible. PhRMA has provided a reasoned reply to the FDA's latest guidelines. The arguments get very technical, but they boil down to using older patients who start off with a higher risk of serious complications. This will assure the FDA that there is an important therapeutic benefit to treatment. PhRMA also proposes to pool the microbiology data across two trials to attain more assurance that we are dealing with bacterial pneumonia while at the same time reducing the total number of patients to be enrolled making the trial more efficient and still valid. In this scenario, if you conduct two trials totaling 1400 patients, you can expect to have a total of about 300–450 with a bacterial diagnosis across both trials. You can examine the results for the new antibiotic and the older antibiotic that is being used for comparison in this smaller subset. Of course, this number of patients will not provide the statistical power that the FDA prefers, but the overall approach will make the trials feasible.

PhRMA notes that enrolling only patients who have had no prior antibiotic therapy will be exceedingly difficult. They propose allowing enrollment of patients who have had less than 24 h of therapy with an antibiotic that has a short half-life (doesn't last long in the blood stream or in the tissues) such that when the patient is started in the trial, the previous antibiotic will have already disappeared from the lungs and should no longer be an important determinant of outcome. Therefore, the study drug will be the most important determinant of the outcome for those patients. The results from the trial can be examined separately for patients who have or have not received any prior antibiotic as a way of assuring the FDA that prior antibiotics made little difference in the trial. If the FDA is unable to see its way clear to acceding to some of PhRMA's suggestions, I don't see that we will have many if any new antibiotics for community-acquired pneumonia for the next decade.

After participating in the latest FDA meeting on clinical trial design in pneumonia, I have offered my own approach to the problem. I think that given a 30–70% treatment effect of antibiotics at day three of treatment of pneumonia, we could easily justify a non-inferiority margin of 15% for the endpoint of clinical improvement at day three. As noted in Chapter 4 where their approach to pneumonia was reviewed, almost no untreated patients with pneumonia were afebrile by day three whereas almost all patients treated with either penicillin or sulfonamides were. If you believe that fever is a sign of illness and resolution of fever is a sign of recovery, which I certainly do, this means that the treatment effect approaches 100%. (I'm exaggerating, it's really only about 80%). Therefore, according the FDA's own formula for calculating a valid margin, that could be as high as 40%. I have proposed a conservative margin of 15%. Then, either allowing the pooling of patients with a microbiological diagnosis across two trials or allowing the use of investigational diagnostic tests to increase the diagnosis rate would make the trial feasible. The only caveat is that the use of such an investigational diagnostic test could not impact the label in any significant way. That is, the company could not be forced to promote use of its antibiotic only under the circumstance where an investigational test was used for the diagnosis. To me, this seems an ultimately reasonable way forward. The key will be to define what "clinical improvement" at day three actually means in practice.

The FDA Has to Stop Moving the Goal Posts in Mid-Stream

The FDA has to get back to keeping its word. Industry cannot and will not continue to invest in expensive clinical trials that are designed in collaboration with the agency only to have the FDA change its mind once the trials are completed and the money has been spent especially when we are talking about a therapeutic area where they view their return on investment to be marginal to begin with. That becomes sunk money for the company and if new trials are required, more money is also required. This is a certain killer for those few biotechs that come up with enough funding for late stage trials. Imagine investors in biotech who work to come up with the $70 million required for late stage trials only to find that after the money has been spent, the FDA has moved the goalposts and wants all or some of the trials re-done. Imagine if they now want investors to spend up to four times the money they initially invested which they now find was wasted. What is to prevent the FDA from moving the goalposts after the next set of trials? And the next?

One company, Replidyne, has already gone belly up because of this practice and others are in the throes of trying to fund additional expensive trials to keep up with the FDA's moving goalposts. The behavior on the part of FDA is simply unacceptable. They have to stick to their word.

The FDA Has to Level the Playing Field with Generic Antibiotics

The FDA cannot continue to allow Americans to use generic drugs that would not meet FDA's modern standards. All drugs have to meet the same standard of efficacy and safety. Specifically, I'm speaking about all the antibiotics that have ever been approved by the FDA for otitis, sinusitis, bronchitis and pneumonia. If the FDA has decided that the trials that were previously carried out do not show benefit, than by definition the risk to benefit ratio for those drugs is zero. Their marketing approval for those indications should be withdrawn. This is especially important since the use of the older generic drugs tends to be much greater than the use of the new antibiotics which is so closely monitored for approval. With the greater use of the older drugs comes the potential for greater danger.

Of course, the better idea would be to rescind the guidelines requiring placebo-controlled trials for these indications and allow new drugs to be developed. As we discussed earlier, placebo-controlled trials are infeasible and sometimes unethical and no one will participate in them – frequently for good reason. If the FDA rescinds their requirement for placebo-controlled trials, the generics can stay on the market and the market can be opened for new antibiotics that we need.

Congress Needs to Step Back from the FDA Especially Where Science Is Concerned

As illustrated by the Ketek fiasco, the FDA must be protected from inappropriate intrusion by congress. Politics has no role to play in the practical and scientific establishment of clinical trial designs. The move by congress to extensively investigate the FDA's handling of antimicrobial clinical trial designs was an unneeded distraction to an already overburdened agency at a critical juncture. FDA staffers were spending hours providing materials for a possible investigation which, in fact, never materialized. It also led to the unnecessary transfer of a very talented and dedicated director of antimicrobial products and ophthalmology that has further eroded FDA leadership in this area. In my view, if the congress has serious scientific questions regarding FDA's conduct, the matter should be turned over to the National Academy of Science that exists for the very purpose of providing neutral scientific assessments for congress. An excellent example of this is the recent report on the FDA and Drug Safety commissioned by congress in the wake of findings of cardiovascular safety issues for Vioxx and other drugs used to treat pain associated with various forms of arthritis. These safety problems ultimately led to the withdrawal of several of these drugs. The National Academy report provides a framework for the establishment of systems to monitor safety of products more carefully at the FDA.

Our Government Needs Access to the Best Advice it Can Get

Both congress and the FDA need to find a way to deal with conflict of interest in a way that allows them access to the best advice. Then, when they get informed opinion, they can weigh the advice in the context of the potential conflicts of those providing it. In my view, as long as there is not a direct conflict with a particular company, experts should be allowed to participate on advisory committees. For example, take someone like me (but not me, really). I consult with a number of small pharmaceutical companies and even have stock in some. But I don't deal with any of the large pharmaceutical companies and would be free of conflict with their products if they ever came before the FDA. The only potential conflict would be if one of my clients were developing a drug similar to the one coming before the committee. The potential conflict in such an example would depend on the definition of "similar." If "similar" means all antibiotics, I would always be conflicted. If "similar" means that the antibiotic would have to have a similar molecular structure and have activity against similar bacteria, then I would only rarely have a conflict. I prefer the latter and use that as the definition for conflict of interest in my consulting business. If such a conflict were to exist, I would have to declare this and recues myself from that particular meeting or portion of the meeting or the FDA would have the option of requesting that I recues myself.

Outside of concerns regarding conflicts of interest, the FDA also should be less constrained in general in choosing people for these advisory committees. It is hard enough to fill these positions with the appropriate expertise. Issues of distribution of race, geography, ethnic background and other characteristics should come after expertise in choosing committee members.

How About a Totally New Approach to Drug Development?

A few years ago, I worked with the Manhattan Institute on providing a way for the FDA to deliver on its own "critical path" initiative where the FDA proposed a new paradigm for drug development. The new paradigm was based on the new genomic science where toxicities and efficacy could potentially be tailored to individual patient-drug combinations if we could know the exact site(s) of action of the drug (which we will probably never really know up front) and the exact genetic nature of the patient or their illness (like cancer). In the field of cancer, this approach is starting to become reality. Unfortunately, the path as spelled out by the FDA was a general wish list with little in the way of regulatory detail. The Manhattan Institute set out to find a specific way forward for the FDA's ideas. Although the report was focused specifically on using molecular medicine and genomics to address problems, part of their proposal could, in my view, be applied to drug development in general. In fact, through the European approach of automatically reassessing and essentially reapproving drugs at regular intervals following the original approval, a part of the Manhattan Institute proposal is already being used.

The Manhattan Institute proposed that a drug be approved conditionally. This approval would be based on trials in some area where there was a clear medical need. An example might be to approve a drug like Vioxx (a pain reliever from Merck that was withdrawn because of heart toxicity a few years ago) only for use in patients with rheumatoid arthritis who had not responded to other pain-relievers. At the same time, while the company was selling the drug to this relatively small population with a high medical need, they would be forbidden to use direct-to-consumer advertising – no TV or magazine advertisements. They could also then be running clinical trials in other indications to identify whether the drug would work in those indications and to identify any toxicity missed in the earlier trials. The company would have to establish a registry to track use and toxicity in the patients being prescribed the drug based on the conditional approval (patients with rheumatoid arthritis where other pain relievers had failed). In this way, the drug would be used in a smaller population that represents a more "real world" population with a true medical need. The population might be in the millions compared to the few hundreds or thousands usually studied in a clinical trial setting. In this way, rare but serious side effects could be detected early and development could either be halted or could be redirected towards those populations where the toxicity might be tolerated. Such populations might include those for which other pain relievers are ineffective, for example. This plan was presented to the FDA back in 2005. The questions that arose were around what the registry would look like and what kind of trials would be required to achieve the sort of conditional approval we discussed. In other words, the FDA did not say "no." We were not laughed out of the room. But I have no idea whether there was any follow-up in the FDA on our presentation.

The idea of conditional approval is now used routinely in Europe. Drugs are approved based on the usual regulatory pathway, most commonly through large, pivotal phase III trials, but the approval is conditional on a review at some future date. If potential risks are perceived in the data presented from the pivotal trials, the conditional approval is short and the company might be asked to provide some risk management plan like a registry to look for safety problems. They might also be requested to carry out additional trials to examine safety while the drug is sold to a more restricted population. If not, the usual review period is 5 years. At the time of review, various approaches are taken. The voluntary adverse events reporting system database might be examined, for example, looking for excess toxicity compared to other drugs used in the same therapeutic area. Or, if a registry had been put in place, those data would be scrutinized. At the time of the review, the population allowed to use the drug might be expanded or the company might fail to garner renewal of marketing authority for the compound based on the data. As I noted above, some variation on this could be applied universally. Everything hinges on the questions raised by the FDA in 2005. What does the registry look like? What do the initial trials look like? In my view, the greater the medical need, the smaller the trials would have to be.

This sort of system seems to be a win-win. Patients with important medical needs could get access to therapy earlier by taking on some additional risk. The pharmaceutical company lowers their own risk by earning money during the period of restricted conditional approval while they explore other indications and obtain additional data on safety. Consumers win because the drug will have been examined in a much larger real world population than would ever have been possible in the usual clinical trial setting. If a safety issue were to emerge, everyone could reflect further on which populations if any might be more appropriate targets for the drug before final approval.

Communication Between the Industry and Physicians Is Required, but is Becoming More and More Restricted

Congress in particular is restricting the ability of the pharmaceutical industry to communicate with physicians and patients. Mostly, this is probably a good thing. For truly innovative products, though, even where there is a clear medical need, there is also an important educational need. In cases like those, how does industry educate physicians about such a new product? Should they only target physicians with educational efforts or should they also try and educate patients? If so, how? These questions are timely given our recent debates over direct to consumer advertisements by the pharmaceutical industry and the righteous indignation of some over connections between the industry and influential academic physicians.

I have had a couple of personal examples of trying to deal with these questions that I would like to share with you. The first came when my team at Wyeth was trying to develop a drug for Respiratory-Syncytial Virus (RSV). RSV causes severe respiratory disease and death in infants and young children and causes everything from colds to pneumonia in older children and adults. RSV is deadly to immunocompromised patients like those who are recovering from bone marrow transplants. For the population as a whole, it is just as deadly as influenza for which we now have both vaccines and antiviral drugs. For RSV, there is only one product marketed. It is an antibody used to prevent disease in premature infants with severe underlying cardiac or lung disease. That's it. One problem is, because there is no treatment, physicians don't try and diagnose the disease. That would mean that if there were a new treatment, no one would know when to use it since they wouldn't know if their patient had RSV, influenza or just an infection with cold virus. If physicians didn't use it, there would be no market. If there were no market, Wyeth (or anyone else) would not develop the drug. One thing we were clearly going to need was going to be ways to diagnose RSV infection reliably in the physician's office similar to the rapid strep tests doctors now use to diagnose strep throat. There we were, stuck in the position of having to develop a market through education of both physicians and patients, even when everyone agreed that there was a clear medical need. Unfortunately, the Wyeth compound turned out to cause birth deformities in animals and was never pursued very far in human trials. So we never had to confront these questions beyond the long-term speculation stage. RSV remains a key medical need and the National Institutes of Health and the Centers for Disease Control have targeted new RSV vaccines for many years. So far, we have nothing other than the antibody I already mentioned.

A second example is from the world of vaccines. At the time I was at Wyeth, their vaccines group was in the midst of large, pivotal clinical trials investigating their new pneumonia vaccine targeting *Streptococcus pneumoniae* and now sold as Prevnar. They targeted the prevention of severe invasive disease –bacteremia (bacteria invading the blood stream) or meningitis (infection of the lining of the brain and spinal cord) with their vaccine. Each year in the United States, pneumococcal disease accounted for an estimated 3000 cases of meningitis and 50,000 cases of bacteremia each associated with high mortality rates. But when Wyeth first started research on the vaccine and they realized it might work, they asked pediatricians and family practitioners how many cases of pneumococcal bacteremia or meningitis they saw in a year and how beneficial the vaccine would be to them in their practice. Very few said they saw more than one or two cases a year and even fewer thought the vaccine would benefit them in their practices. Of course, many academic pediatricians working in large referral centers saw significant numbers of these cases and realized immediately how important a product this could be. In collaboration with a number of these academic clinicians and investigators, Wyeth then embarked on an educational campaign to show clinicians the importance of this disease on a more global basis. By the time the vaccine was launched, the vast majority of these same practitioners agreed that this vaccine would be an important benefit to both them and their patients. Among children under age 5, 7 years after introduction of the vaccine, invasive pneumococcal disease has decreased by 45% overall and the decrease has been 94% if you just look at the specific strains targeted by the vaccine. There has even been a benefit to non-vaccinated children and adults probably because fewer children are carrying disease-causing strains of bacteria following vaccination. Numerous studies have now demonstrated that this vaccine saves lives and is cost-effective. I wonder whether in today's environment such an educational campaign could actually be carried out and if not, what effect the lack of such a campaign might have had on this life-saving vaccine. Would Wyeth have ever developed the vaccine? Would it have received a recommendation for routine use by the Centers for Disease Control (CDC)? Would the uptake of the vaccine and therefore the benefit of the vaccine have been as dramatic even if both of those conditions were fulfilled?

Again, a balanced approach is required here. A good mix of openness and transparency on the part of both industry and physicians would allow everyone to see potential conflicts of interest and to judge for themselves as to whether the information that is being provided is trustworthy or not. But to portray physicians and others who work with industry as somehow biased or worse, just plain corrupted from the outset, will not help us obtain the new therapies that we so desperately need.

What Should the Industry Be Doing?

Finally, the pharmaceutical industry itself can provide a solution. The story of Roy Vagelos and his efforts while CEO of Merck to bring Ivermectin, a drug discovered by Merck as useful for heartworm in pet animals, to patients in central Africa afflicted with river blindness is an inspiring one. It turns out that Ivermectin was active against the parasite, *Onchocerca volvulus* that causes river blindness. It is transmitted by the bites of flies. Vagelos championed this cause within Merck and engineered a collaboration with the World Health Organization to carry out clinical trials and eventually to distribute the drug. The drug was ultimately approved and Merck has donated it to everyone in need of this therapy. The drug has saved the sight of countless numbers of Africans and other affected populations around the world. This story has served as an inspiration to many of us working in the industry for many reasons. Vagelos, a physician by training, came up to the position of CEO from that of a researcher at Merck. He understood the good that could come from important therapies and he understood human medical needs. There are now many examples throughout the pharmaceutical industry of this sort of approach to diseases in the developing and in the developed world going from trachoma, another cause of blindness, to malaria to HIV and AIDS to TB.

When a researcher from Johnson and Johnson was asked why that company was remaining committed to antibiotics discovery research (they have now essentially abandoned their effort), they replied that their CEO felt a strong sense of social responsibility. In fact, Johnson and Johnson, through their Tibotec subsidiary, is even now collaborating with a number of non-profits to bring a new TB therapy to market.

One way forward for the industry, then, is for them to develop a new and urgent sense of social responsibility in the area of antibiotic resistance. This may, however, be a tough sell to their shareholders because of the costs involved. For antibiotic resistance, there might not be the ready availability of the World Health Organization to help organize and carry out appropriate trials, for example.

The Current Large PhRMA Model Is not Viable. These Giants Need to Divide Themselves into Bite-Size Portions to Survive

The industry, in my view, cannot continue to survive in its current large company blockbuster-dependent configuration. There are simply not enough multi-billion dollar drugs being discovered to support all those companies that need them for survival. In Chapter 6, I noted the position of Pfizer once they will have completed their acquisition of Wyeth. They will be a $60 billion company needing to generate anywhere from $3 to 6 billion in additional revenues (or cost savings) every year. There will be cost savings after the merger, but eventually Pfizer will have to come up with one or more very big products to survive. And this is not just true of Pfizer, but applies to all of the current giants with 10s of billions of dollars in revenue coming in each year.

As one way of achieving this growth, the industry has recently invested heavily in cancer research. 7000 of 22,000 clinical trials for drugs listed on the PhRMA website are for cancer therapy. Industry is attracted to a field where small gains like an extra 6 weeks of life, can garner high prices and large profits. They are also attracted to it because they have a good opportunity to make the diagnoses much more specific by identifying which specific mutations exist in each person's cancer so they can provide personalized therapy for that cancer. In that way, they would be able to target a tiny fraction of cancer patients, but achieve a much higher success rate. This will still allow very high prices, but I'm not so sure about profits since the drug could only be used by a very small number of patients.

I believe that they only way forward financially for large pharma, completely ignoring the problem of antibiotics for the moment, is for them to deconsolidate. The large pharmaceutical companies have become slow moving behemoths. They are now bastions of bureaucracy which rival large governments in the time it takes them to make and implement decisions. The company that seems to be the farthest along this road is Glaxo Smith Kline, but they have not yet made the final two leaps. A number of years ago, GSK reorganized into CEDDs or centers of excellence in drug discovery. The CEDDs are in turn organized around therapeutic areas. They currently have five CEDDs one of which is in infectious diseases, encompassing antiviral and antibiotic drug discovery and early development. Each CEDD is responsible for development through Phase II, the stage that normally leads directly to the pivotal Phase III trials that are so expensive. A centralized clinical development group within GSK handles these later trials. But in terms of profit and loss, it is GSK as a whole that reports to Wall Street, not the CEDDs individually. So GSK is still responsible for putting everything together to make up the increase in revenues required to attract additional investors. GSK still holds major control over the overall company portfolio, so the independence of the CEDDs is limited. If, for example, there are potential blockbuster drugs coming from other therapeutic areas, the infectious diseases CEDD might be last in line for phase III or manufacturing resources from the centralized GSK resource. Or they might not be given sufficient budget monies to contract these tasks independently of the GSK central resources. The same would be true for in-licensing a new drug from biotech. This sort of decision is made on a more company-wide basis where projects are prioritized.

GSK has innovated another approach that comes even closer to the one I would prefer. They have formed a joint venture with Pfizer for the development and marketing of antiviral drugs targeting HIV/AIDS. This HIV NEWCO as they call it can cherry pick drug candidates from the Pfizer or the GSK discovery research effort, or they can license drugs from the biotech in a relatively independent fashion. The NEWCO has considerable independence from both Pfizer and GSK central decision-making. But, like all other therapeutic areas, they do not report their results directly to the markets, but rather these would be captured by the parent companies.

The deconsolidated model I would propose would place responsibility for profit and loss within each of the therapeutic areas. Ideally, each area would report separately to Wall Street and the investors. In the deconsolidated model, then, the revenues from all of GSK's branded infectious diseases products, antibiotics and antivirals, would flow back to the infectious diseases therapeutic area minus some "tax" for the umbrella organization or holding company. Each therapeutic area could then garner its own investments through a stock listing, private funds or other options. They would be responsible for their own budget from drug discovery through marketing. Each area would also have to duplicate manufacturing and late stage development to one extent or another. The larger umbrella company, GSK, would extract a fee or tax – a percentage of revenue – from each therapeutic area. With this money, which would have to be returned to the various therapeutic areas under some sort of prioritization scheme, the umbrella organization would make centralized decisions about which therapeutic areas require additional support. This support could provide some insurance against downturns for one of the areas, or could allow some promising programs to be accelerated. The larger umbrella organization could offer certain centralized services such as phase III clinical development, commercial scale chemical manufacturing and marketing. But in this case, they would bid like any contract manufacturer for the business of any of the therapeutic areas. I believe that a central marketing service would be important since it would allow synergy from marketing products from multiple therapeutic areas with a single sales force. A central resource for the large Phase III trials would also be advantageous. In a model where each therapeutic area would have its own profit and loss control and its own budget entirely, decisions about funding large phase III trials, about manufacturing and about in-licensing compounds from biotech would be internal to the therapeutic area as long as the resulting cost fell within their budget. Their budget, in general, would depend on revenues from their products. Although I don't know the exact figures for revenues from infectious diseases products at GSK, I would guess they would come in at several billion dollars. That means that revenues for anti-infectives at GSK would only have to increase by several hundred million dollars to satisfy Wall Street's demand for investment growth.

The small organization would be better poised to carry out its responsibility to provide growth for shareholders since it would be more nimble and quick.

It would be better able to deal with the evolving regulatory standards in the sense of being able to adapt more rapidly. Of course, if the trials required become so expensive that the deconsolidated entities could not afford to do them, we are all sunk anyway. Because the scale of the company would be so much smaller, the normally slow uptake of antibiotics in the marketplace would be easier to handle compared to the rapid uptake demanded by giant pharmaceutical companies.

Arguments in favor of a strong, centralized organization are many. They include the greater negotiating power of a large, multi-therapeutic area company in dealing with contractors. A large company can also achieve synergies by allowing central clinical development groups to manage trials in multiple therapeutic areas within smaller countries such as those in South America or places like South Korea or Taiwan. Personally, I think the advantages of deconsolidation outweigh any of the advantages of the strong, centralized bureaucracy, but obviously not everyone agrees with me.

The other option that would be good for antibiotics, but that would not solve large pharma's problem, would be simply to spin off the antibiotics business entirely from the large pharma company. So, lets take the new Pfizer-Wyeth example. In terms of antibiotics, the combined company will have; piperacillin-tazobactam, a billion dollar product that is just starting to feel major effects from loss of its patent position; tigecycline, launched in 2005 and now selling around $250 million; and linezolid which is selling over $1 billion but which is threatened with loss of its patent exclusivity in 2015. For a start-up company, this would be a very strong beginning. It would allow a rapid entry into the public markets for additional investors. The new company could then go about its business of both trying to discover and develop its own new antibiotics and to bring in products from outside, perhaps from biotech, to fill out its pipeline. In my view, the new company would be much smaller than its current configuration within the larger company in the sense that it would have a small number of key internal folks who would manage programs through, mainly, outside contractors much as is done in biotech. The marketing organization could be more streamlined and could incorporate marketing collaborations with external parties. I'm not sure that such a structure is more expensive than using internal large pharma services and it does allow for a good deal more flexibility. This solution also has the advantage of avoiding the enormous large pharma bureaucracy, allowing the new company to react more quickly to changing situations and to make decisions more efficiently.

Chapter 8
Conclusions

It is clear to me that the millennia-old ongoing war between antibiotics and microorganisms will continue regardless of anything man does to intervene. In other words, all of our efforts to get care-givers to wash their hands between patients, to get hospitals to clean the environment properly and to convince physicians to use antibiotics only when necessary will be too little too late. Of course, I don't mean to belittle these efforts that I think are an important part of our armamentarium to preserve the utility of the antibiotics we have. But to deal with the uncanny ability of organisms to become resistant and to spread the genes that determine resistance to all their cousins around the globe, we simply need a constant stream of new antibiotics. True, eventually, with time, the microorganisms will become resistant to the new antibiotics, which is why we need to keep finding new ones. It is and will always be never-ending just as it has been since the beginning of life on earth.

You would think that once we, as a global society, realized the truth of this incontrovertible scientific conclusion, we would find a way to make sure that we fulfilled our responsibilities to provide for this clear medical need. But nothing is logical and action does not necessarily follow from truth. The forces that conspire against the discovery, development and marketing of new antibiotics are formidable. First, those that seem the easiest to control, our own regulatory agencies, continue to develop policies that will only further discourage the development of new antibiotics. This by itself has far-reaching consequences that most don't realize. If the regulatory environment is such that development becomes expensive or simply infeasible, as is becoming the case for antibiotics in the US and even, though to a lesser extent, in Europe, companies give up. Of course, companies give up for lots of other reasons as well. The science of discovery is getting harder and harder. The market is less and less attractive. If companies drop out of antibiotic research, this trickles down to our academic centers where there is less incentive to train people for careers in antibiotics and where there is less likelihood for academics to partner with industry for innovative products. So the academics do what everyone else does – they go where the money is – heart disease, cancer, arthritis, anything but antibiotics.

D.M. Shlaes, *Antibiotics*, DOI 10.1007/978-90-481-9057-7_8,

Of course, to a certain extent, the NIH can reverse this trend to obsolescence of scientific training in antibiotics by offering grant money to support such training in academic centers. But what would the motivation for the NIH to do this be if there were no way forward to make new antibiotics available to the public?

Biotech is affected even more than academia. Biotech, mostly, remains dependent on large pharmaceutical companies to develop and market their products. This is especially true as the products approach the latest and most expensive stages of development or, when the products are ready to be launched onto the marketplace. Biotech usually cannot afford to pay for the large, late stage pivotal clinical trials required to obtain marketing approval nor can they afford to amass a sales force with enough critical mass to launch a new product. But as large pharma drops out of antibiotic research, they lose the internal expertise required to even evaluate opportunities that might come from biotech. So they simply don't try. Of the seven large pharmaceutical companies that have dropped out of antibiotics research, only one has been actively engaged in licensing antibiotics from biotech. And that company, J&J, only just recently cut their antibiotics research group. So it is not clear that they will continue to license antibiotics from biotech in the future.

With the loss of a critical mass of researchers in both industry and in academia, innovation pays the price. But to get around organisms that are already resistant to everything we have, innovation is what we need. It is sad for me to note that there is very little innovation in antibiotics occurring either in biotech, in large pharma or in academia. Of the antibiotics in late stage development, very few actually originated either in academia or in biotech. The vast majority are cast-offs from large pharmaceutical companies developed by biotech and later, sometimes, re-acquired by large PhRMA.

To halt our inexorable slide into a pre-antibiotic era, it is clear that a multipronged approach is required. The first effort must come from the regulatory agencies to provide a reasonable pathway for companies to obtain approval for antibiotics in indications where there is still some market potential. Insisting on infeasible trial designs for approval takes us in the wrong direction.

Government must do its share to provide some incentive to companies to stay in an area with a high medical need but where the market opportunity is marginal. Two proposals here would probably work. The wild-card patent exclusivity whereby a company would be allowed an additional 6–24 months of patent exclusivity on a product from its portfolio of the company's choice if they successfully bring to market an antibiotic active against the key ESKAPE pathogens is a strategy that would be successful. As I noted in Chapter 7, some of the best and most profound thinking on the antibiotics problem is coming from Europe. And, if Europe is able to implement its key proposals, they will lead the way forward for all of us. The "push-pull" proposal from Europe, in my view, could be successful if both the push and the pull were of sufficient size. In this strategy, the government provides support for late stages of clinical development as a push. If the needed antibiotic is approved, the government exercises its option and purchases a certain predetermined quantity of the antibiotic at a predetermined price at launch and perhaps for several years thereafter.

The establishment of the Transatlantic Taskforce on antimicrobial resistance is a hopeful sign that government finally realizes the seriousness of the problem we are facing. The question remains whether they will be able to act in a significant way to stem the tide and whether their actions will come in a timely manner.

Finally, industry itself will have to change. In my view, breaking the large companies down into much smaller and more facile parts. This could be achieved in two ways. The large company could break itself down into small, independent business units. These units could either be privately held or public companies. In this way, the anti-infectives unit would be obligated to increase revenues according to its own size and not according to the needs of the large pharmaceutical giant to which it now belongs. A second option for antibiotics would be for the larger companies to simply spin out a small, independent company. The spin-off could either be privately or publicly held. This would relieve antibiotics from the burden of trying to fit within the blockbuster needs of the giant parent firm.

I see the size and extent of the storm threatening our miracle drugs, antibiotics. I see how we can protect ourselves and how we can make sure that we never relapse into a pre-antibiotic era for patients and their physicians. I am encouraged by the fact that others such as the Infectious Diseases Society of America, Sweden, the Council of the European Commission and the Obama administration see it as well. I can only hope that appropriate action is taken before it is too late.

Bibliography

Bad bugs no drugs (2004) A monograph from the Infectious Diseases Society of America. 2004. http://www.idsociety.org/BBNDWhitePaper04.htm. Accessed 20 Aug 2010

Boucher HW, Talbot GH, Bradley JS, Edwards JE, Gilbert D, Rice LB, Scheld M, Spellberg B, Bartlett J (2009) Bad bugs, no drugs: no ESKAPE! An update from the Infectious Diseases Society of America. Clin Infect Dis 48:1–12

Concept paper on the need for revision of the note for guidance on evaluation of medicinal products indicated for treatment of bacterial infections (CPMP/EWP/558/95 REV 1) (2009). http://www.ema.europa.eu/pdfs/human/ewp/43563508en.pdf. Accessed 20 Aug 2010

DiMasi JA (2001) Risks in new drug development: approval success rates for investigational drugs. Clin Pharmacol Ther 69:297–307

DiMasi JA, Grabowski HG, Vernon J (2004) R&D costs and returns by therapeutic category. Drug Info J 38:211–223

Duran GM, Marshall DL (2005) Ready-to-eat shrimp as an international vehicle of antibiotic-resistant bacteria. J Food Prot 68(11):2395–2401

Elemam A, Rahimian J, Mandell W (2009) Infection with Panresistant Klebsiella pneumoniae: a report of 2 cases and a brief review of the literature. Clin Infect Dis 49:271–274

Gonzales R, Bartlett JG, Besser RE, Cooper RJ, Hickner JM, Hoffman JR, Sande MA (2001) Practicve guideline. Principles of appropriate antibiotic use for treatment of acute respiratory tract infections in adults. Ann Intern Med 134:479–486

Graham D, Warner WP, Dauphinee JA, Dickson RC (1939) The Treatment Of Pneumococcal Pneumonia with Dagenan. CMAJ 40:325–32

Gwaltney JM Jr, Wiesinger BA, Patrie JT (2004) Acute community-acquired bacterial sinusitis: the value of antimicrobial treatment and the natural history. Clin Infect Dis 38:227–233

Hidron AI, Edwards JR, Patel J, Horan TC, Sievert DM, Pollock DA, Fridkin SK (2008) NHSN annual update: antimicrobial-resistant pathogens associated with healthcare-associated infections: annual summary of data reported to the National Healthcare Safety Network at the Centers for Disease Control and Prevention, 2006–2007. Infect Control Hosp Epidemiol 29:996–1011

Innovative Incentives for Effective Antibacterials. http://www.se2009.eu/polopoly_fs/1.25861!menu/standard/file/Antibacterials5.pdf. Accessed 20 Aug 2010

Innovative incentives for effective antibiotics – Adoption of Council conclusions [Public debate pursuant to Article 8(3) CRP (proposed by the Presidency)]. http://register.consilium.europa.eu/pdf/en/09/st16/st16006.en09.pdf. Accessed 20 Aug 2010

Lax E (2004) The mold in Dr. Florey's coat. Henry Holt, New York

Lederberg J et al (1998) Antimicrobial resistance. A report of a workshop held by the forum on emerging infections of the Institute of Medicine. National Academy Press, Washington, DC

Lord Soulsby of Swaffham Prior (2008) The 2008 Garrod Lecture: Antimicrobial resistance—animals and the environment. Journal of Antimicrob Chemother 62:229–233

D.M. Shlaes, *Antibiotics*, DOI 10.1007/978-90-481-9057-7,

Mataseje LF, Neumann N, Crago B, Baudry P, Zhanel GG, Louie M, Mulvey MR (2009) ARO water study group. Characterization of cefoxitin-resistant Escherichia coli isolates from recreational beaches and private drinking water in Canada between 2004 and 2006. Antimicrob Agents Chemother 53:3126–3130

McCormick DP, Chonmaitree T, Pittman C, Saeed K, Friedman NR, Uchida T, Baldwin CD (2005) Nonsevere acute otitis media: A clinical trial comparing outcomes of watchful waiting versus immediate antibiotic treatment. Pediatrics 115:1455–1465

McDonald LC, Owings M, Jernigan DB (2006) Clostridium difficile infection in patients discharged from US short-stay hospitals, 1996–2003. Emerg Infect Dis 12:409–415

Moran GJ, Krishnadasan A, Gorwitz RJ, Fosheim GE, McDougal LK, Carey RB, Talan DA (2006) Methicillin-resistant S. aureus infections among patients in the emergency department. N Engl J Med 355(7):666–674

Mossialos E, Morel C, Edwards S, Berenson J, Gemmill-Toyama M, Brogan D Policies and incentives for promoting innovation in antibiotic research. http://www2.lse.ac.uk/LSEHealthAndSocialCare/LSEHealth/News/Antibiotics%20Report.aspx. Accessed 20 Aug 2010

Nordmann P, Cuzon G, Naas T (2009) The real threat of Klebsiella pneumoniae carbapenemase-producing bacteria. Lancet Infect Dis 9:228–36

O'Shea JG (1990) "Two minutes with venus, two years with mercury" – mercury as an antisyphilitic chemotherapeutic agent. J Royal Soc Med 83:392–395

Payne DJ, Gwynn MN, Holmes DJ, Pompliano DL (2007) Drugs for bad bugs: confronting the challenges of antibacterial discovery. Nat Rev Drug Discov 6:29–40

Ram FS, Rodriguez-Roisin R, Granados-Navarrete A, Garcia-Aymerich J, Barnes NC (2006) Antibiotics for exacerbations of chronic obstructive pulmonary disease. Cochrane Database Syst Rev Apr 19;(2):CD004403

Shlaes DM, Moellering RC (2008) Telithromycin and the FDA: implications for the future. http://infection.thelancet.com Vol 8 February 2008

Shlaes DM, Moellering RC Jr (2002) The United States Food and Drug Administration and the end of antibiotics. CID 2002:34:420–422

Shlaes DM, Projan SJ (2008) Antimicrobial resistance vs. the discovery and development of new antimicrobials. In: Meyers DL (ed) Antimicrobial drug resistance. Humana Press, New York, pp 43–50

Shlaes DM, Projan SJ, Edwards JE Jr (2004) Antibiotic discovery: State of the state. ASM News 70:275–281

Snow V, Mottur-Pilson C, Hickner JM (2001) For the American College of Physicians–American Society of Internal Medicine. Principles of Appropriate Antibiotic Use for Acute Sinusitis in Adults. Ann Intern Med 134:495–497

Spellberg B, Miller LG, Kuo MN, Bradley J, Scheld WM, Edwards JE Jr (2007) Societal costs versus savings from wild-card patent extension legislation to spur critically needed antibiotic development. Infection 35:167–174

Spellberg B, Talbot GH, Boucher HW, Bradley JS, Gilbert D, Scheld WM, Edwards J Jr, Bartlett JG (2009) Antimicrobial availability task force of the infectious diseases society of America. Antimicrobial agents for complicated skin and skin-structure infections: justification of noninferiority margins in the absence of placebo-controlled trials. Clin Infect Dis Aug 1; 49(3):383–391

Spellberg B, Talbot GH, Brass EP, Bradley JS, Boucher HW, Gilbert DN (2008) Infectious Diseases Society of America. Position paper: recommended design features of future clinical trials of antibacterial agents for community-acquired pneumonia. Clin Infect Dis 47(Suppl 3): S249–265

Spellberg B, Talbot GH, Brass EP, Bradley JS, Boucher HW, Gilbert DN (2008) Infectious Diseases Society of America. Position paper: recommended design features of future clinical trials of antibacterial agents for community-acquired pneumonia. Clin Infect Dis Dec 1;47(Suppl 3):S249–265

The Bacterial Challenge, a Time to React, published by the European Centre for Disease Prevention and Control and the European Medicines Agency. http://www.ecdc.europa.eu/en/publications/Publications/0909_TER_The_Bacterial_Challenge_Time_to_React.pdf

Thornurn AL (1983) Paul Ehrlich: pioneer of chemotherapy and cure by arsenic (1854–1915). Br J Vener Dis 59:404–405

Time Magazine. Dec. 28, 1936. Medicine: Prontosil

Vernacchio L, Vezina RM, Mitchell AA (2007) Management of acute otitis media by primary care physicians: trends since the release of the 2004 American Academy of Pediatrics/American Academy of Family Physicians Clinical Practice Guideline. Pediatrics 120:281–287

Welte T, Petermann W, Schurmann D, Bauer TT, Reimnitz P, and the MOXIRAPID Study Group (2005) Treatment with sequential intravenous or oral moxifloxacin was associated with faster clinical improvement than was standard therapy for hospitalized patients with community-acquired Pneumonia who received initial parenteral therapy. Clin Infect Dis 41:1697–1705

Zhanel GG, Hisanaga TL, Laing NM, DeCorby MR, Nichol KA, Palatnik LP, Johnson J, Noreddin A, Harding GK, Nicolle LE, Hoban DJ (2005) NAUTICA Group. Antibiotic resistance in outpatient urinary isolates: final results from the North American urinary tract infection collaborative alliance (NAUTICA). Int J Antimicrob Agents 26:380–388

Index